T0280890

Cambridge Elements ≡

Elements in the Philosophy of Science
edited by
Jacob Stegenga
University of Cambridge

PHILOSOPHY OF OPEN SCIENCE

Sabina Leonelli
University of Exeter

CAMBRIDGE
UNIVERSITY PRESS

CAMBRIDGE
UNIVERSITY PRESS

Shaftesbury Road, Cambridge CB2 8EA, United Kingdom

One Liberty Plaza, 20th Floor, New York, NY 10006, USA

477 Williamstown Road, Port Melbourne, VIC 3207, Australia

314–321, 3rd Floor, Plot 3, Splendor Forum, Jasola District Centre,
New Delhi – 110025, India

103 Penang Road, #05–06/07, Visioncrest Commercial, Singapore 238467

Cambridge University Press is part of Cambridge University Press & Assessment,
a department of the University of Cambridge.

We share the University's mission to contribute to society through the pursuit of
education, learning and research at the highest international levels of excellence.

www.cambridge.org
Information on this title: www.cambridge.org/9781009416399

DOI: 10.1017/9781009416368

© Sabina Leonelli 2023

This work is in copyright. It is subject to statutory exceptions and to the provisions
of relevant licensing agreements; with the exception of the Creative Commons version
the link for which is provided below, no reproduction of any part of this work may take
place without the written permission of Cambridge University Press & Assessment.

An online version of this work is published at doi.org/10.1017/9781009416368 under
a Creative Commons Open Access license CC-BY-NC 4.0 which permits re-use,
distribution and reproduction in any medium for non-commercial purposes providing
appropriate credit to the original work is given and any changes made are indicated.
To view a copy of this license visit https://creativecommons.org/licenses/by-nc/4.0

All versions of this work may contain content reproduced under license from third
parties.

Permission to reproduce this third-party content must be obtained from these
third-parties directly.

When citing this work, please include a reference to the DOI 10.1017/9781009416368

First published 2023

A catalogue record for this publication is available from the British Library.

ISBN 978-1-009-41639-9 Paperback
ISSN 2517-7273 (online)
ISSN 2517-7265 (print)

Cambridge University Press & Assessment has no responsibility for the persistence
or accuracy of URLs for external or third-party internet websites referred to in this
publication and does not guarantee that any content on such websites is, or will
remain, accurate or appropriate.

Philosophy of Open Science

Elements in the Philosophy of Science

DOI: 10.1017/9781009416368
First published online: July 2023

Sabina Leonelli
University of Exeter

Author for correspondence: Sabina Leonelli, s.leonelli@exeter.ac.uk

Abstract: The Open Science (OS) movement aims to foster the wide dissemination, scrutiny and reuse of research components for the good of science and society. This Element examines the role played by OS principles and practices within contemporary research and how this relates to the epistemology of science. After reviewing some of the concerns that have prompted calls for more openness, it highlights how the interpretation of openness as the sharing of resources, so often encountered in OS initiatives and policies, may have the unwanted effect of constraining epistemic diversity and worsening epistemic injustice, resulting in unreliable and unethical scientific knowledge. By contrast, this Element proposes to frame openness as the effort to establish judicious connections among systems of practice, predicated on a process-oriented view of research as a tool for effective and responsible agency. This title is also available as Open Access on Cambridge Core.

Keywords: research practice, equity in research, diversity in research, Open Society, science policy

ISBNs: 9781009416399 (PB), 9781009416368 (OC)
ISSNs: 2517-7273 (online), 2517-7265 (print)

Contents

1 Introduction

Openness has long been a guiding principle for liberal democracies, where recognition of the epistemic significance of transparent, free and inclusive inquiry is a source of both political and scientific legitimacy. Just as politicians owe their credibility and influence to their perceived accountability vis-à-vis the electorate, scientists owe their credibility and influence to the perceived effectiveness and breadth of the scrutiny applied to their research. Openness is often viewed as a necessary complement to accountability and public scrutiny. As argued by philosophers ranging from Karl Popper to Jürgen Habermas, Helen Longino and Philip Kitcher, what distinguishes a dictator from an elected leader – or a scientist from a crook – is the extent to which their decision-making processes are visible, intelligible and receptive to critique.

The Open Science (OS) movement, with its emphasis on ensuring that research outputs, components and methods are widely disseminated, scrutinized and reused for the good of science and society, is but the latest chapter in the historical co-evolution of political and scientific accountability. In this sense, the movement is neither novel nor surprising, and maintains a strong continuity with values long viewed as definitive of scientific research – such as the critical questioning of dogmas, the search for reliable evidence, the privileging of rational reasoning and the emphasis on public scrutiny and debate. At the same time, OS has gathered momentum over the last three decades as a response to the broad transformations brought about by the digitalization, globalization and commodification of research. As new technologies and an ever-growing workforce massively increase the volume and velocity of discoveries, questions around what constitutes effective communication become more urgent, with scientific institutions struggling to adapt their practices to the collaborative exigencies of the contemporary world. Insofar as it strives to respond to these developments, OS is all about novelty: it is explicitly geared towards transforming the research system as currently construed, thus potentially revolutionizing the ways in which the scientific process is construed, performed and assessed.

A key component of this transformation is a renewed attention to the multiplicity and diversity of outputs produced over the course of scientific inquiry. Open Science is widely portrayed as an opportunity to redesign research practices, evaluation and governance to better highlight and utilize such outputs, including books and articles but also data, models, software, techniques, instruments, samples and other research constituents whose epistemic value has arguably been underestimated within science communication and credit systems. Hence the blossoming of digital infrastructures to guarantee free and instant access to research papers, data and models ('Open Access', 'Open Data', 'Open Methods');

standardized note-taking tools, such as digital lab books, to help document and eventually replicate research procedures ('Open Notebooks'); reviewing systems that, rather than looking for original tive contributions to existing knowledge, assess the robustness and validity of research outputs, thus fostering publication of all high-quality results without necessarily making assumptions around what may be especially significant and for whom ('Open Peer Review'); and collaborative venues to foster the exchange of insights and materials across national, disciplinary, professional and cultural borders – particularly through forms of public engagement that bring insights from non-scientists into research ('Citizen/Community Science'). Public and private institutions around the world have set up strategies to support OS initiatives, ranging from national roadmaps to international treaties, online publishing platforms, updated checks on research quality and revised metrics for scholarly excellence. Politicians have also embraced OS with renewed vigour, presenting it as an effective mechanism to transform basic research into 'scientific capital' for future innovation,[1] and thereby reasserting the deep link between the political and scientific roles of openness. From corporate boardrooms to university management and political positioning, debate over the significance of OS and its implementation has risen to the top of the agenda.

This Element presents a philosophically informed reading of the epistemic role of OS within contemporary research: how OS policies and practices affect research methods and outputs, what this means for the nature and structure of scientific inquiry, and how the very idea of openness can and should be understood in relation to the pursuit of knowledge about the world. This is not meant as a purely descriptive take on current OS practices, though long-term engagement with those practices, as briefly discussed below, strongly inform my views. Rather, this Element presents a normative interpretation of the history, motivations and potential of OS, focusing on broad trends characterizing its current implementation. My aim is to provide a constructively critical reading of the commitment to transparency and sharing often made within the OS movement, which has in my view become an obstacle to the movement's efforts to promote reliable and responsible research. I argue that one step towards addressing this concern is the adoption of a different philosophical standpoint, one where openness is conceptualized not as primarily about sharing resources but rather as primarily fostering meaningful communication between the humans involved in research. Making this broad argument requires me, unavoidably, to provide a general characterization of the OS movement that does little justice to its complexity and multiplicity. Let me thus state this

---

[1] A long-standing twentieth-century agenda in science policy, as pursued by Vannevar Bush in the wake of World War II.

upfront: this Element does not mean to capture the vast and diverse landscape of OS initiatives in any comprehensive way, and there are many realities within OS that do in fact abide by the understanding of openness as connection which I am partial to. Nevertheless, my analysis captures discourse and commitments that are frequently found especially in large-scale OS initiatives and policies, which in my view deserve critical discussion. Hence this Element builds on empirical research around the history and current functioning of OS, yet provides an interpretation of such materials that is explicitly grounded in a normative perspective.

This approach is reflected – and inspired by – an understanding of ethics as integral to epistemology in the tradition of standpoint theory and strong objectivity (Harding 1995), whereby one's perspective on a subject is always a 'view from somewhere' coloured by one's background and goals. My overall interest in this Element is to support the future development of OS by providing a philosophical framework for what openness could and should mean for research aimed at sustaining life on this planet. I am specifically interested in the use of OS to pursue the public good, including to enrich existing understandings of what forms such 'good' may take depending on publics and contexts.[2] In keeping with this overall philosophical stance, I shall consistently intertwine epistemic and ethical considerations as grounding for my analysis of research practices. As I shall illustrate, ethical concerns around the discriminatory and exclusionary implications of some OS practices are impossible to disentangle from epistemic concerns around the reliability and robustness of research produced through those practices. The methodological soundness of procedures of sampling, representation, modelling, communication and interpretation depends on both technical features and social context.[3]

Historically, my starting point is two complementary observations. First is the radical significance of pursuing openness in research at the time of writing, when the hopes raised in the 1980s by the rise of the World Wide Web and related communication technologies are giving way to disillusionment at the widespread deployment of digital tools to curtail, obfuscate or misdirect the free circulation and critical scrutiny of ideas. Despite the illustrious history of openness as the cornerstone of liberal thinking, the 2020s are not a time for naïve calls for 'openness for its own sake', whatever that may mean. As the Internet becomes a playground for corporate monopoly and fake news threatens to overwhelm attempts at earnest debate, the dangers and misuses of the idea of free information have become apparent for all to see. This has severe

---

[2] This stance builds upon like-minded views of Longino (1990), Kitcher (2001), Wylie (2003), Rouse (1987, 2015), Potochnik (2017), Cartwright et al. (2022) and Chang (2022), among others.
[3] Beaulieu and Leonelli (2021) and Thompson (2022).

implications for the way in which openness is conceptualized and enacted in relation to scientific research.

The second observation is that, despite the good intentions and the vast efforts committed to their actualization, OS initiatives are fraught with difficulties and are sometimes met with resistance by the very research communities that they are meant to serve. This observation is corroborated by a growing body of international scholarship centred on OS implementation, including extensive qualitative research that I carried out over the last decade, in collaboration with colleagues across the natural and social sciences, to investigate how researchers across countries and domains perceive OS and its implications for their work. We found that in contexts where researchers receive relevant support and training, OS can increase the quality and inclusivity of scientific debate. However, the vast majority of researchers work in disciplines and institutions that are not internationally visible, well-funded and/or attuned to rewarding OS efforts. This makes it difficult for them to use OS infrastructures to support their work, since the design of those infrastructures reflects the interests, assumptions, priorities, skills and technological resources of their developers – who are often English-speaking scientists based in rich institutions where such work can be supported.[4] I have complemented such work with research documenting the history of ideas of openness and collaboration across the sciences, as well as personal involvement in large-scale efforts by various research and policy organizations to identify conditions under which OS could be actualized.[5] Through such experiences I witnessed considerable disagreement over what OS involves and what roles openness and transparency play in knowledge production and use.[6] It is from consideration of the roots and implications of

---

[4] Leonelli (2016), Levin et al. (2016), Bezuidenhout et al. (2017), Chen et al. (2019), Leonelli (2022a), Ross-Hellauer et al. (2022). Another prominent source of worry among researchers is the exploitation of OS by commercial entities (part of broader trends towards digital feudalism: Jensen 2020) and organizations interested in distorting scientific results for political reasons (e.g. debates over climate change: Lewandowsky and Bishop 2016, Nerlich et al. 2018).

[5] My forays into science policy stemmed from research conducted since 2007 on the epistemology of big data, which highlighted the significance of novel ways to mobilize and reuse data towards transforming science. Requests to report on such research led to participation in numerous debates around Open Data, Open Access and OS infrastructure; and roles as researchers' representative or expert advisor for the Global Young Academy, the European Commission, Plan S and the International Council of Science, among others. The resulting reports are available on the Open Science Studies website (www.opensciencestudies.eu); see also Burgelman (2021), Miedema (2021) and Owen et al. (2021) for insider reflections on academic involvement in these policy debates.

[6] While largely built on the study of scientific practices in biology and biomedicine, my analysis is meant to also embrace the social sciences and humanities, whose perspectives I have learnt about through interaction with social scientists and colleagues in philosophy, history and literature studies, and through advisory roles in research organizations overseeing social science and humanities portfolios. Given this ample remit, throughout the text I use the term science in the

such frictions, rather than from the polished statements associated with the political call to 'open up science', that my analysis departs.

A crucial problem is lack of clarity over how OS, with its emphasis on multiplying research avenues, outputs and participants, relates to the existing diversity in epistemic practices utilized by different research communities around the globe – and, in turn, to the varying socio-political settings in which research takes place. It is widely recognized that operating in an OS landscape requires effective communication, which in turn demands some level of consensus around common procedures, standards, principles and metrics. In other words, making decisions around how to open science unavoidably involves deciding what may and may not count as 'good' science;[7] and insofar as OS infrastructures can function as sources of reliable knowledge, they can also act as tools to identify and police questionable research practices. In response to these requirements, many of the more institutionalized OS initiatives tend to privilege a homogenous, universally applicable understanding of the scientific method over a pluralistic and situated one. It is much easier to set up OS guidelines when assuming that science consists of a coherent body of knowledge and procedures that can and should conform to common norms – an assumption that flies in the face of the rampant plurality of research approaches used across domains, locations and contexts, and the significance of such plurality in delivering a robust, comprehensive and reliable understanding of the world.[8]

As yet, there is little systematic understanding of how openness relates to the standards and criteria of best practice developed and performed by researchers around the world to suit their specific goals and working conditions. In what follows, I argue that in the absence of such understanding, the high level of standardization and precise validation practices demanded by some OS initiatives threatens to blindly privilege specific ways of knowing, thus potentially disrupting sophisticated methodologies, inadvertently dismissing well-established research traditions, and exacerbating the already large epistemic and social divides separating research domains and locations. As denounced by a number of critics in science and science studies, there is a substantive risk of

---

continental sense of *Wissenschaft*, comprising humanities as well as the social and natural sciences.

[7] This is also why it is impossible to keep a rigid distinction between discussions of OS and discussions of science as a whole: in this Element, the focus on OS often and unavoidably expands to embrace broader debates around what research looks like in the twenty-first century, and what this means for future science practice and policy.

[8] There is an enormous body of scholarship on scientific pluralism, which I cannot hope to comprehensively review in this Element. I focus on salient aspects, predominantly extracted from the philosophy of science, in Section 4.

some OS policies – despite their good intentions and progressive slant – acting as a reactionary force which reinforces conservatism, discrimination, commodification and inequality in research, thus ultimately closing down opportunities for inquiry in a disastrous reversal of what they set out to achieve. I maintain that it is possible to rescue OS from such a fate by highlighting OS initiatives grounded on a deep understanding of local knowledges and their social context, and that an important step in that direction is to articulate which understanding of scientific practice – in other words, which *philosophy of science* – best underpins the goals set by the OS movement. This is what this Element aspires to contribute, starting from an analysis of the roots, motivations and implications of interpreting openness as anchored on the sharing of research components, and then arguing for an alternative view centred on the reticular and distributed development of research processes, as already exemplified by many grassroots OS projects which consistently engage with the interests, preferences and methods underpinning specific ways of knowing.

The argument is set out in four sections (Sections 2–5). Section 2 reviews some *key features of the contemporary OS movement*, focusing on systemic problems plaguing the global research landscape – and particularly existing constraints on research communication, collaboration and publishing – and OS attempts to address such problems through an expansion of what counts as research output and the provision of incentives to share such outputs as widely as possible. I argue that underpinning many such initiatives is a vision of openness as the *freedom to share* resources and insights at various stages of the research process, whereby the adoption of incentives towards making results more transparent is expected to increase the reproducibility and accessibility of research, leading to more inclusive, engaged and reliable forms of inquiry. In principle, this vision of OS seems unassailable, an effective reaction to a scientific system that has become increasingly opaque, exclusive and commodified. The question that concerns me, however, is how this vision plays out within actual research settings.

Section 3 confronts this question by shifting the analytic focus from the theory to *the practice of OS within everyday scientific work*. I briefly examine four examples of OS implementation, including: (1) the effort to share biological data on the SARS-CoV-2 virus responsible for the coronavirus pandemic, which has been widely hailed as a demonstration of the effectiveness of OS in fostering discovery under emergency conditions; (2) current challenges to the evaluation of quality standards for data, models and software, and the extent to which such evaluation depends on tacit assumptions about which technologies may enhance or even guarantee data validity; (3) the development of global infrastructures to link locally sourced data about crops and their environments, which is critical to research on food security and planetary health, yet is conditioned by pre-existing inequities between data producers and users; and (4) the use of specific interpretations of the notion of reproducibility as a criterion for what may constitute reliable research methods. These examples illustrate, on the one hand, how the tremendous diversity in goals, values, targets, background knowledge and material settings within contemporary science results in different expectations around best practice; and, on the other hand, how such diversity can be squashed by demands for fast and smooth sharing of scientific resources, which can damage scientific advancement while also failing to address the systemic problems discussed in Section 2.

Having explored one interpretation of openness and found it wanting in research practice, the next step is to explore alternative interpretations of openness that may take better account of scientific diversity and the empirical insights garnered from studies of how researchers conduct, communicate and discuss their work. To this aim, Section 4 builds on philosophical literature on scientific pluralism to *identify four central characteristics of systems of research practice,* which in my view need to be acknowledged and supported by OS initiatives: (1) specificity to local conditions; (2) entrenchment within research repertoires; (3) permeability to newcomers; and (4) demarcation strategies. From this analysis I conclude that it is impossible to foster or even evaluate the quality of scientific procedures and outputs without considering how research conditions change across locations, who is included and excluded from specific ways of conducting research, and with what implications for the structure of inquiry and the knowledge being produced. I end by discussing the interrelations between *epistemic diversity* and *epistemic injustice,* arguing that both play a crucial role in the development of good science, and need to be placed at the centre of OS initiatives.

The analysis of OS practices presented in Sections 3 and 4 allows me, in Section 5, to expand my critique and sketch an alternative vision that better underpins the quest for reliable and responsible research practices. This requires digging further into the epistemic foundations of the idea of openness as sharing. I argue that this view is entangled with an *object-oriented framing of the epistemology of science* as a matter of control over resources, where questions around which forms of expertise are brought to bear on the research process remain secondary to the production of tangible outputs and the development of standard procedures and agreements over how to trade such outputs and thereby accrue their value. Within such framing, science is construed as consisting in the accumulation of facts, methods and insights, whose free circulation, scaffolded by technologically sophisticated infrastructures, suffices to guarantee research progress as well as the opportunity for different parts of society to deploy those resources towards addressing urgent challenges. I contend that this view of research is misleading and unrealistic, and that related understandings of openness are unlikely to deliver the epistemic benefits associated with the OS movement in the long term. This is not because the technologically mediated sharing of resources is not relevant to scientific development, but rather because sharing does not constitute a necessary starting point nor a sufficient condition for conducting reliable and responsible OS. As an alternative, I propose a conception of *openness as judicious connection*, which is grounded in a *process-oriented epistemology of science* that recognizes the situated, embodied and goal-directed nature of communication and collaboration among researchers. This understanding of openness emphasizes the dynamics of science as a human enterprise that brings different ways of acting and understanding the world in relation with each other, and thus fosters many different forms of output selection, organization and interpretation. Under this interpretation, Open Access is not achieved solely by making access to publications free of charge, but rather by fostering publication on the basis of fair assessment of its quality and irrespectively of authors' ability to pay for processing charges; Open Methods is not a matter of recording and sharing every detail of a research procedure, but rather a reflection on which research components and techniques are most salient to the outcomes, and should thus be accessible and reproducible; Citizen Science does not involve offloading labour-intensive parts of data collection to participants without involving them into the research process, but rather building relationships with non-professional publics who bring relevant insight; and Open Data does not mean the sheer accumulation of research data on digital platforms, but rather the recognition that not all data can or should be made available, and choices need to be made and justified around which data are being shared, and how data infrastructures may support the creative exploration of such data.

This framing of OS takes epistemic diversity and justice as guiding principles for producing reliable knowledge. Open Science initiatives need to question explicitly and regularly what is considered a scientific contribution, for which purposes and by whom. This means recognizing that effective sharing is built on well-justified, contextualized discrimination and judgement over the value and goals of research and its components, rather than absence of judgement, disregard for the specificity of research conditions and related attempts to 'make everything available'. Scientific discovery is thus positioned as a social and situated endeavour, thereby underscoring the links between OS, existing understandings of good practice, and specific conceptions of what an Open Society may look like.

2 Rethinking Communication: Research in a Changing World

This section discusses the core motivations and constituents of OS. After identifying problems afflicting contemporary research, I describe the emergence of the OS movement as means to tackle these problems. I conclude with an analysis of the values underpinning the OS landscape, emphasizing the prominent role played by the interpretation of openness as the freedom to share resources and ideas.

2.1 Research Troubles: The Dominance of 'Closed' Science

A starting point for any discussion of openness in research is questioning what is perceived as 'closed', and why. Indeed, one way to frame the OS movement is as a reaction to current forms of scientific communication, and particularly a culture of research publishing that is ever more competitive, commercialized and self-referential. The last fifty years have witnessed an explosion in the scale and international reach of research efforts, with increasing numbers of people training as professional scientists across the globe. Accordingly, research outputs have grown exponentially and have become ever more specialized, putting pressure on existing systems to disseminate and evaluate findings; and publishing services have grown in technological and administrative sophistication, fostering a vast ecosystem of specialized journals and indexing tools to help readers wade through the deluge of information. While researchers have contributed free labour as authors, reviewers and editors, financial support for publishing has come from ever-higher subscription charges, which eventually have made access to academic journals unaffordable to all but wealthy research institutions. At the same time, intense competition for jobs and grants, with growing numbers of applicants and shrinking percentages of success, has put pressure on the systems of assessment used to determine who produces good research and deserves employment and funding. In many countries and research institutions, the quantity of papers produced and the prestige of the journal in which they are published have become a shorthand for research quality and reliability.

This situation is exemplified by the popularity of the impact factor, which quantifies how many times articles in each journal have been cited in a given period. Originally meant as a measure of the quality of journals, rather than of single contributions or contributors, the impact factor has been widely adopted to gauge individual authors' influence on their research domain. Use of this metric strengthened the stronghold acquired by publishers on research communications and evaluation. For most systems of scientific assessment around the world, papers published in high impact factor journals continue to be the only recognized output of the research process, with little regard for other components – such as data, models, software and instruments – viewed as mere means to the authoring of an article.[9] Thus the pursuit of knowledge is imagined as the ordered assemblage of objects and procedures which, like modular building blocks, scaffold the writing of texts; and knowledge itself has become commodified into article-shaped units (somewhat ironically referred to as 'minimal publishable') whose production is often subsidized through public funding or public-private partnerships, but which are only available to those who can afford subscription fees.[10] This generates high profits for publishing companies and greatly limits the recognized forms and potential publics of scientific knowledge. And this concerns only cases where publishing actually happens – an important qualification since vast swathes of scientific results are never released outside the institution that generates them. These include much of the science carried out under military or industrial funding, which is sometimes so secretive that it is not even known what *topics* are being investigated at any one time (let alone where, by whom and with what results). It can also include publicly funded research on sensitive subjects, whose outputs may be withheld for a variety of reasons including data protection or concern over the potential implications of the findings.

The commercialization of scientific communication and its reduction to journal publications have created profound epistemic issues (Radder 2010, 2019). This system does not offer incentives or rewards for the responsible dissemination and scrutiny of research components other than whatever is presented as part of a publication. Most research data, models and procedures are not circulated beyond the group of researchers who generated them; even when they are published (on a database, for instance, or as supplementary information to a research paper), there is little institutional recognition for the extensive efforts involved in making such components available for scrutiny. The focus on publishing a high quantity of articles encourages researchers to publish their results as fast as possible, which is hard to combine with thorough checking procedures, replicating experiments, validating results through a variety of methods, formatting data for wider dissemination, or consulting relevant experts from other domains who may help to improve research design and contextualize outputs (Edwards and Roy 2017). Moreover, just as the pressure to win funding leads researchers to inflate the potential of their work, the pressure to publish in prestigious journals results in a tendency to overstate

---

[9] Patents may also be accorded high status as desirable outputs, though even they are not easy to incorporate into quantitative metrics focused on the impact factor.

[10] I discuss the problematic nature of this modularized, object-oriented epistemology of science in Section 5.

and/or overgeneralize one's results, thereby making assumptions about the scope and validity of inferences that are not necessarily well justified. One would expect such problems, which include statistical cheats such as p-hacking and publication bias, to be identified and addressed during the peer review process. However, reviewing is itself not a visible or rewarded activity despite its central role in publication, making it difficult to find volunteers for this laborious task and ensure that reviews are as thorough as they should be (Heesen and Bright 2019). There is also little formal training for reviewers, with early career researchers typically thrown into that role without any preparation except for scant guidelines by publishing venues, their own idiosyncratic experience as recipients of referees' comments, and equally idiosyncratic advice from senior academic mentors.

These trends raise concern around the trustworthiness of published research and the effectiveness of current scientific communication. Such concerns are exacerbated when considering the implications of this system for early career researchers, who need a high number of publications to secure a research job and yet are responsible for the most labour- and time-intensive research activities, such as data gathering. The emphasis on publication quantity is training researchers to skimp on detail and rigour in favour of hype and speed. It is also discouraging them from diversifying their methods: for instance, by complementing quantitative research with qualitative findings or vice versa, which is a rewarding but time-consuming effort; and by collaborating with those outside their immediate specialized networks, whose diverse viewpoints and expertise may not fit discipline-focused forms of assessment. The future generation is therefore being pushed away from transdisciplinary research and robust investigative practices. The chances to rebel are small, given that publication-obsessed cultures privilege those who have long held academic jobs and lack incentives to address prejudice, ageism, bullying, misogyny and racism. Furthermore, the large influence of some governmental and industrial funders in picking worthy social and environmental challenges leaves scientists with little autonomy over their research agenda and publishing goals.

Of course such trouble in science relates to broader social issues, including the lack of diversity among elite, specialized jobs; a persistent Eurocentric bias in the conceptualization and assessment of relevant forms of expertise; deepening political and socio-economic conflict, which impedes transnational collaboration and the opportunity to include multiple perspectives on global problems; a short-term understanding of the scientific, political and economic benefits of research, which is exemplified by the projectification of science into units to be completed within an average of three to five years, and discourages long-term investment in research infrastructures and venues; and the increasing weaponization of scientific authority by groups with vested interest, ranging from ultra-right campaigners to corporate public relations, whose ability to mimic empirical methods to legitimize or disqualify claims has further increased since the advent of social media. The continuing success of authoritarian regimes and the threat of armed conflict are strengthening nationalist trends that disrupt scientific collaborations across borders (Krige 2022), while powerful private actors – such as the fossil fuel industry and big tech corporations – improve their ability to colonize scientific debate to their advantage (Oreskes and Conway 2010). The Russian invasion of Ukraine and the subsequent boycott of Russian science by Western institutions are only one recent instance of these trends. In the meantime an aggressive market economy, increasingly costly infrastructures and shifting perceptions of the relation between research and education are threatening the financial viability of many research endeavours, while individual investigators buckle under a proliferation of demands and skills including funding, management, media presence, policy engagement and technological prowess (not to speak of the actual research). Taken together, these issues represent an existential threat to research institutions.

2.2 Looking for a Solution: The Open Science Movement

It is hard to come away from such a depiction of scientific knowledge production without worrying that the whole research system is bankrupt (Allison et al. 2016). This impression is reinforced by debates around the so-called 'reproducibility crisis', building on widely publicized failures to replicate seminal experiments in psychology and biomedicine – as well as the willingness of some political and corporate actors to point at cracks in the evidence used to justify policy, as recently instantiated by polemics around COVID-19 vaccines and prevention measures. What can be done to improve the situation? Many researchers, activists and policy makers working within the OS movement see the inaccessibility of research – whether through paywalled publications or through unpublished data – as a central issue. A red thread running through the list of scientific woes is the perceived inscrutability and unaccountability of research – a pernicious form of 'closure' that stands in the way of engagement, understanding and feedback. Hence the insistence on opening up research as a panacea to redress the wrongs, rescue science from commercial and political exploitation, and bring it back to its core values.

What such core values may be has long been a subject of debate within the history of science, with dramatic shifts in the moral and epistemic discourse around what attitudes and practices may foster reliable research (e.g. Rothblatt 1985). Among the steadiest moral and epistemic commitments within Western science is the emphasis on *sharing resources*, whether they be samples, instruments, texts or measurements, from which knowledge can be extracted. The invitation to support research through sharing has taken countless forms through the ages: from the ideal of the universal library that made Nineveh, Babylon and Alexandria into capitals of the ancient world, to the medieval obsession with collecting rare objects and observations, the large-scale circulation of knowledge and goods overseen by modern colonial empires, and the information networks envisaged by neo-liberal economists as the foundation of free-market capitalism. Such sharing has not been, for the most part, for the benefit of a broad public; nor has it always been fair or transparent, given its strong association with violent forms of conquest and appropriation, and a markedly non-democratic understanding of the goals and beneficiaries of scientific inquiry. Nevertheless, the emphasis on sharing resources did underscore the distributed nature of learning and the relevance of mobilizing research outputs for reuse across contexts and locations. To this day, sharing continues to be construed as the starting point for scientific investigation – most famously through the Baconian view of empirical knowledge as a fundamentally cumulative endeavour, grounded in the skilled gathering of facts which appropriately trained humans can interpret to derive or discard hypotheses.

The sophisticated view of inductive reasoning underpinning this approach to empirical knowledge points to another core trait of science in addition to sharing, which is the provision of *critical feedback* by individuals with relevant skills and background. Much of the institutionalization of Western science during the Scientific Revolution, emblematically represented by the Royal Society of London and its journal *Philosophical Transactions*, consisted in developing specialized venues for such exchanges – venues where both the collected materials and their analysis would be subject to the scrutiny of peers, and budding researchers would be socialized into questioning received views. In the early 1940s sociologist Robert Merton famously identified such 'organized skepticism' as a defining feature of science vis-à-vis other human endeavors, alongside the norm of 'communism' which emphasizes equal access to resources that may be required to contribute to scientific efforts (Merton 1942).

This immediately raises the question of what constitutes acceptable forms of feedback and who is recognized as possessing the skills, knowledge and resources required to perform critique and thus participate in the advancement of scientific knowledge. This question has garnered significance as science became professionalized as a plausible – if still elitist – career choice beyond the ranks of wealthy male patrons and their protégées, making the identification and assessment of relevant expertise into a foundational issue for the scientific enterprise. Merton sought to address this issue by emphasizing the role of scientific institutions in demarcating scientifically relevant expertise from the personal experiences and values of researchers. He thus proposed two add-itional norms for science, each emphasizing the conditions under which indi-vidual judgements within scientific collectives may be deemed reliable: universalism, which encourages a disregard for the particularities of each researcher's social background and experiences in favour of an impersonal view of scientific knowledge; and disinterestedness, the commitment to divest expert feedback of vested interests. Merton's emphasis on the role of institu-tions in demarcating relevant from irrelevant expertise found its complement in Karl Popper's (1945) defence of what he called 'Open Society'. This work, which explicitly introduced the term 'openness' to scientific and political debate, takes rational deliberation within the sciences as a model for liberal society, and similarly points to the relevance of institutions in drawing bound-aries between acceptable and unacceptable interventions. In Popper's view, democratic institutions play the fundamental role of establishing ground rules for what constitutes rational arguments and credible evidence, while also encouraging debate over whether and how such rules should be modified and improved.

Through Popper's framing, openness was formalized as individuals' freedom to exercise judgement and reflect on the very conditions under which such judgement is evaluated. This understanding of openness had a significant influ-ence on post-war political debates in Europe and North America, with their tight association of democratic and scientific orders. Yet, as discussed in the previous section, it was less effective in shaping the organization and institutional imagin-ation of science, arguably due to the strong emphasis on the role of individual agency within an imagined free marketplace of ideas over the role of collective agency within a highly unequal and mostly unregulated institutional landscape. Mertonian concerns with equitable and disinterested access to information were overwhelmed by successful forms of knowledge commodification by powerful, well-resourced players, making the idea of 'open exchange' into a contradiction in terms. Merton's and Popper's attention to the social conditions for deliberation and critique were arguably pushed to the background by market forces and the pursuit of financial value increasingly permeating academic institutions. What was left was the emphasis on sharing and accumulation as key conditions for knowledge, and the related view that science was the tradeable product of specific configurations of information control (Leonelli 2019a).

It is thus no surprise that the first stirrings of the OS movement as we know it today started in the late 1970s with critical discussions of the dangers of obstructing access to knowledge and the scientific, social and economic benefits derived from the free sharing and scrutiny of resources (Chubin 1985). The development of open source software in the 1980s, which emerged as a counterpoint to corporate attempts to commodify code, played an important role in solidifying this understanding of openness as *freedom to share*, with programmers and participants in hacker culture exemplifying effective collaboration to produce freely accessible tools such as, eventually, Apache and Linux (Kelty 2007). By the time the expression 'open science' made its appearance in print in 1985, it was unavoidably tied to a rebellion against commercial and legal strictures imposed on access to and participation in scientific efforts, and an emphasis on novel information and communication technologies – further fuelled by the open source release of the World Wide Web in 1993 – as crucial tools to overcome these barriers.

In the following two decades, the OS ecosystem came to encompass several intersecting initiatives aiming to liberate research outputs, broadly conceived, from the ownership claims that so fundamentally structure the research system (Bartling and Friesike 2014; Nerlich et al. 2018; Montgomery et al. 2021; Miedema 2021). Key among them were data and preprint repositories built to bypass corporate controls over research content and enable anybody to view and utilize outputs for their own purposes. In 1991 came the creation of ArXiv, a research-sharing platform initially aimed at releasing articles in advance of formal publication in a journal, which quickly became a model for Open Access. The subsequent rise of Open Access journals such as PLOS was accompanied by an emphasis on publishing datasets alongside research findings, which for fields such as climate science and molecular biology dated back to early twentieth-century efforts to set up dedicated data infrastructures on global weather patterns (Edwards 2010) or organisms of interest (Leonelli and Ankeny 2012). The rise of large data infrastructures was further strengthened in the 2010s by the emergence of data journals and data release policies by major funders and scientific organizations. This push for Open Data mirrored the growing scientific and financial value ascribed to data as scientific outputs in and of themselves, and proved more dramatic than Open Access in the ways it challenged existing understandings of what constitutes a publishable output and, relatedly, research labour and credit structures (especially given that data generation and stewardship are typically viewed as low-status occupations for technicians and students rather than as high-status contributions to discovery; Leonelli 2016). This in turn led to complementary calls for Open Materials, Instruments, Models, Methods, Notebooks (including lab books, field notes and other ways of describing research techniques and procedures) and even Open Education (in the form of training materials teaching, among other things, how to use OS tools). The impenetrability of peer review processes was also questioned, with proposals to make feedback and revisions to scientific findings more transparent by publishing both reviews and responses online (Fecher and Friesike 2014). In the same vein, the OS movement incorporated a push towards citizen science (Hecker et al. 2018), and particularly its promise to increase participation in research through activities such as the crowdsourcing of data collection and analysis – which jointly epitomize the emphasis on accumulating and sharing resources through engagement with an ever-growing workforce (Strasser et al. 2018; Prainsack 2020).

It is important to note that, despite their initial impetus, OS efforts to share resources are not always framed in opposition to the commercialization of research outputs, and some parts of OS have in fact become increasingly aligned with the existing economy of publishing. Developing and implementing OS tools and procedures costs money and expertise, raising serious questions as to what business models and financial sources may support it. Best placed in this respect are the largest commercial publishers, who are perfectly positioned to corner what for them is effectively a new market. The publishing giants Springer-Nature, Taylor and Francis, Elsevier, SAGE and Black & Wiley have indeed enthusiastically endorsed the OS movement – correctly interpreting it as a phenomenon that could further expand their remit and reinforce their centrality to research efforts (Mirowski 2018). An obvious instance was the launch of the 'author-pays' model of Open Access, through which commercial publishers recoup the money lost from journal subscriptions by charging authors whose papers are accepted for publication. Many of the largest publishing houses also successfully deployed their financial and technical resources towards supplying metrics, indexing services and data storage capabilities for OS, thus reinforcing their dominion over research communications.

Looking beyond the academic realm, many of the public and private institutions supporting OS are motivated by the desire to fuel innovation and economic growth. The statements made by organizations ranging from public funders to national governments and philanthropic foundations (such as the European Commission, the Gates Foundation and the Dutch government, among many others) leave no doubt that commercial innovation is among the favoured outcomes of the free use of research components facilitated by OS. Accordingly, OS practices are encouraged particularly within publicly funded research, with the expectation that downstream application of the resulting solutions may well be patented and commercialized for exclusive use of specific providers (Leonelli and Lewandowsky 2023). This promise of increased commodification is often coupled with an emphasis on transparent information sharing as a mechanism to improve democratic governance, whereby voters understand the reasoning and evidence underpinning decision-making, resulting in higher accountability for public institutions (Herzog 2023). Better access to insights produced across the public and (in principle) private sectors is expected to power faster, more effective and socially responsible processes of discovery, leading to innovation that may benefit society as well as the economy. OS is thus interwoven with initiatives around Open Innovation as well as Open Government, which consolidates the image of scientific research as at once legitimizing source and ideal model for rational political and economic interventions within liberal democracies. This interpretation of Popper's Open Society is tightly associated with elective participation in the knowledge economy and related liberal politics and markets.

2.3 Openness as Sharing: From Transparency to Inclusion

My reading of the development and context of the OS movement is opinionated and, again, by no means comprehensive: it is intended to highlight the variety of motivations and expectations therein, as well as the extent to which these motivations converge towards a common understanding of openness as the free and immediate sharing of resources. Open Science stakeholders may differ in their views of who should be envisaged as sharing with whom, and what constitutes preferable or even acceptable usage of the research being shared. Nevertheless, many of them agree on the importance of a seamless flow of research materials, resources and information – and the significance of challenging communication barriers. Openness thus construed revolves around the idea of unlimited access: the desire to make any research element available to anyone who may wish to use it as soon as possible after its creation. Given the potential of novel technologies to collect and instantaneously mobilize outputs, the means envisaged for such access are largely digital. The scale of OS in this interpretation is necessarily global, embracing anybody with an interest in research no matter where they are located – a sentiment emphasized by the willingness to invest in educational materials that could foster skilful engagement. As a result, it is often assumed that OS will have unequivocally good outcomes: it will improve the content and reliability of scientific knowledge as well as researchers' working conditions, thus proving to be good for science and society; and increase equality of participation in research by making previously inaccessible resources available to those who may wish to use them, and making it possible for anybody to scrutinize the evidence and reasoning underpinning scientific claims.

The definition of OS provided by the European Commission in 2015, when openness was formally placed at the heart of EU science policy, illustrates how

Table 1 The main features of
the interpretation of openness
as freedom to share resources.

Openness as sharing
Unlimited
Digital
Good
Global
Equal

the idea of openness as sharing informs the construal of an OS ecosystem. The position paper 'Open Innovation, Open Science, Open to the World' explicitly takes the digitalization of economic transactions as a model for how research should be conducted in the future. Open Science is defined as

> a new approach to the scientific process based on cooperative work and new ways of diffusing knowledge by using digital technologies and new collaborative tools ... sharing and using all available knowledge at an earlier stage in the research process. OS is to science what Web 2.0 was to social and economic transactions: allowing end users to be producers of ideas, relations and services and in doing so enabling new working models, new social relationships and leading to a new modus operandi for science. (European Commission 2016)

Here the mention of 'all available knowledge' blatantly exemplifies the idea of unlimited access and extends it across the entirety of the scientific process. As in many other such policy documents, research is portrayed as encompassing a succession of stages – typically going from research design and data collection to analysis and publication – each of which generates outputs worth sharing. The understanding of OS as a 'new approach' tied to 'new ways of diffusing knowledge' and 'new collaborative tools' underscores the foundational role of digitalization and related technologies to the potential and novelty of OS. Last but not least, the reference to shifting modes of participation in science is explicitly linked to a repositioning of knowledge 'users' into 'consumers' (Radder 2010). This repositioning aims to fill the gulf between academic and non-academic research, thereby – it is assumed – facilitating a more equitable distribution of resources, and more participative and inclusive ways of generating knowledge beyond the strictures and narrow-mindedness associated with the idea of academia as Ivory Tower. Similar goals are expressed in the aptly named report 'Open Science by Design' published in 2018 by the US National

Figure 1 Core values in OS implementation: the current direction of travel.

Academies of Sciences, Engineering, and Medicine, which continues this political trajectory by zooming onto the mechanisms – technologies, investments and institutions – needed by researchers to effectively share resources.

There is a specific direction of travel here, a choreography and prioritization of specific values as the best path towards openness, which arguably underpins many mainstream efforts to implement OS (see Figure 1). First, one needs to achieve *transparency*.[11] This is often presented as the most immediate and urgent preoccupation for OS: the push to put everything online, thereby making it accessible to a wide variety of potential users. Second, one worries about the *quality* of what is being shared. Enter criteria and mechanisms for assessing the reliability of outputs and methods circulated on the web, ideally accompanied by sanctions for those who do not abide by such rules. The notion of reproducibility has garnered enormous attention as precisely one such criteria, whose application across domains is expected to enhance the rigour and trustworthiness of what is being shared online (National Academies 2019). Third, there is *inclusion*, intended as the opportunity for anybody with relevant interests and expertise to engage with and participate in research, and thus to utilize – and help scrutinize – the resources being shared. In most OS policy documents, the end goal and ultimate outcome of improving the transparency and quality (often cashed out as reproducibility) of research is an inclusive and fair research process, which fosters scientific engagement while also helping to discriminate between good and bad contributions to knowledge (e.g. European Commission 2016, 2018; National Academies 2018; Burgelman et al. 2019; United Nations 2019).

The Mertonian and Popperian emphasis on reflexive, critical and institutionally mediated confrontation among individuals has not disappeared from OS thus construed, but it is predicated on the idea of transparency. Only research elements that are widely visible and accessible can be appropriately evaluated as more or less reliable building blocks for future research. Within this view, it is from sharing resources and outputs that desirable outcomes such as legitimacy, participation and

---

[11] For an analysis of various meanings associated with transparency in science, see Elliott (2020).

trust in science – as well as economic growth – can emerge. As I explore in the next section, however, this 'transparency-first' view of sharing does not fare well when considering how OS is implemented, and with what outcomes.

3 Rethinking Practice: Challenges of Open Science Implementation

Section 2 clarified why OS, grounded on a view of openness as sharing, has proved an enticing prospect for researchers, institutions and funders alike. This section considers whether and how such prospects are realized when OS is implemented within everyday scientific practice. I briefly examine four attempts at OS implementation, each of which illustrates opportunities and challenges inherent to opening the research ecosystem. Open Data figure prominently in my choice of examples, since debates around data sharing exemplify core assumptions and expectations around openness held by practicing researchers, as well as the deep link between conceptual and ethical commitments underpinning different visions of how data can be used to generate knowledge. Indeed data management is dependent on, and significantly affects, Open Source, Open Access, Open Peer Review, Open Methods, Open Instruments and Citizen Science. Hence the tensions illustrated by these examples go well beyond the use and governance of data alone: they are intended to provide a window on the intertwined technical, social, ethical and conceptual challenges underpinning efforts to develop and apply OS strategies.

3.1 The Access Wars: COVID-19 Data Sharing

My first case concerns the global sharing of research concerning the COVID-19 pandemic, which has been widely hailed as a demonstration of the value and power of OS towards accelerating research and informing emergency measures such as social distancing, quarantines and the development of vaccines. The dissemination of genomic data about the coronavirus SARS-CoV-2 has been particularly successful, with many discoveries – including the identification of new variants of interest, the mechanisms underpinning COVID vaccines and potential avenues to treat severe infections – resulting from the opportunity to swiftly share data on virus strains across hundreds of research sites around the world. And yet, some data sharing initiatives in this space have come under a barrage of attacks for 'not being open enough' and for posing 'barriers that restrain effective data sharing'.[12] One reason for the attack was due to the fact that, contrary to existing covenants within molecular biology to share non-human

---

[12] Open Letter (2021), subsequently reported in *Nature*.

genetic data without constraints (Maxson Jones et al. 2018), some of these initiatives posed limits on how the data could be accessed and reused.

Let us briefly consider the case of the Global Initiative on Sharing All Influenza Data (GISAID). This database, originally created in 2008 to share data on influenza, was swiftly redeployed in January 2020 to provide access to SARS-CoV-2 data. It requires its users to sign an agreement, which includes assurances about adequately crediting the original data producers and constrains how GISAID-stored data may be linked and integrated with other sources. This requirement stems from the recognition that some researchers – often working in low-resourced environments and/or less visible research locations – are reluctant to share data due to fears of better-equipped researchers building on such work without due acknowledgment. Such fears are justified. Reusing data available online requires reliable and powerful connectivity and computing resources, as well as the adoption of standards that match the theoretical perspectives and material capabilities of laboratories based in rich institutions. Hence researchers based in low-resourced environments cannot always take advantage of Open Data, no matter how innovative and rigorous their work may be, and remain reluctant to contribute their own data to online collections (Bezuidenhout et al. 2017). The GISAID user policy is a crude but relatively effective attempt to offset these problems. Having a formal agreement and credit structure in place has fostered information exchange among groups that differ considerably in their geo-political locations, funding levels, material resources and social characteristics – with researchers from 240 countries sharing a total of 15 million sequences by February 2023 (https://gisaid.org). At the same time, the requirement to account for the extent to which data can be accessed and linked limits the integration of GISAID data with other sources, thus negatively affecting the pace and breadth of research – leading to backlash by hundreds of leading researchers concerned about the urgency of an effective pandemic response.

The Global Initiative on Sharing All Influenza Data was built on the recognition of entrenched differences in power, resources and visibility among research groups. Its data governance structure – strongly focused on safeguarding the rights of data producers – is an attempt to counter inequity between researchers based in the high-resourced and those in the low-resourced institutions. In choosing to sidestep these issues, researchers calling for 'fully open' data are prioritizing the idea of transparency over questions of inclusion. This attitude is understandable: the ability to bulk download and freely explore/link COVID data facilitates novel observations and multiple interpretations of those resources. Such advantages cannot be underestimated, especially at a time when researchers are scrambling for data resources that are large and well-constructed enough to facilitate immediate analysis and interpretation. And yet, the focus on the technical and scientific advantages of unlimited sharing takes attention away from the sociocultural factors (such as the geo-political location and characteristics of researchers), institutional issues (such as power dynamics among research sites and expectations around intellectual property) and infrastructural resources (such as the availability of funding and dependable connectivity) underpinning data reuse. Ignoring such factors carries an epistemic risk. It puts a premium on fast-paced research scaffolded by high-tech infrastructure – factors which, however, are not always or necessarily marks of quality and long-term reliability (Chen et al. 2019, Leonelli 2021). It can result in the exclusion from data sharing initiatives of researchers who are not based at prestigious academic institutions, which in turn reduces the diversity and range of data available online, as well as the types of expertise viewed as significant when evaluating that evidence base (Sheehan et al. 2023). These are significant ethical issues in terms of who is included and recognized as a participant in research; but far from being solely an ethical problem, this form of bias has also epistemic implications, since it substantially reduces both the diversity of data sources available to biomedical researchers globally and the amount of expertise – and particularly localized, regional expertise – put to the service of data modelling and interpretation. While there is good reason to critique GISAID procedures as unhelpful to large-scale data mining and analysis, GISAID attention to the contexts within which data are generated, credited and commodified constitutes a valuable attempt to underscore the *scientific* implications of inequity within research and develop solutions that improve the evidence base for future investigations.

The debate over GISAID and its governance exemplifies how efforts to abide by the principle of openness, particularly when openness is interpreted first and foremost as a form of sharing, can clash with responsible research measures geared towards protecting researchers whose work is unrecognized and/or discriminated against. Science is not a level playing field. Providing trustworthy and explicitly non-exploitative conditions for data dissemination helps widen participation in research, which in turn expands the evidence base for subsequent discoveries (Chen et al. 2019). It can also help prevent the circulation of low-quality data (Leonelli 2018a), the widening of digital divides (Bezuidenhout et al. 2017) and the pursuit of socially harmful research (Elliott and Resnik 2019). Rather than attacking GISAID as an example of bad OS, it would therefore seem more fruitful to help improve the usability of GISAID data, while also recognizing the limits of fully transparent, uncredited data dissemination.

3.2 The Mangle of Practice: Technology and Quality Standards

Despite the rise of China and India as scientific superpowers and the vertiginous growth of the scientific workforce in Africa, South East Asia and South America – all of whom challenged the supremacy of Western science in setting universal standards for rationality and best practice (Harding 2011) – regimes of assessment, credit and quality control set up by privileged institutions modelled on Anglo-American academia continue to rule research rankings and evaluation regimes. One consequence is a widespread perception, by researchers themselves and by those who evaluate their work, that good research practice depends at least in part on access to specific technologies. This perception has significant implications for OS efforts.

Consider the importance assigned by OS to free open-source software (FOSS), which can be accessed and modified without constraints or expensive fees, thus posing no barriers to its adoption – contrary to the expensive subscriptions and controls characterizing proprietary software. The adoption of FOSS for scientific research may seem uncontroversially beneficial to those who have little financial backing, until one looks at how researchers in low-resourced environments select and use software to support their work. A survey carried out by the Global Young Academy among researchers in Bangladesh, Ghana and Tanzania, for instance, highlighted a preference for using expensive proprietary software (Vermeir et al. 2018). This was confirmed even in cases where equivalent FOSS alternatives were available and obtaining funds to pay for proprietary tools was difficult if not impossible. One reason for this preference was the perceived stigma attached to using open software. Some participants in the study thought that editors and referees of international journals would interpret FOSS use as a mark of low-quality research, particularly when coming from research locations with little international reputation. Using well-recognized proprietary software such as *MatLab* and *Mathematica*, by contrast, was seen to align with global expectations around appropriate methodology, thus facilitating publication of research results by Anglo-American journals. Similar arguments have been made around quality assessment for datasets, which is often understood to depend on the technology used to produce the data – with the latest models of high-throughput genome sequencers, for instance, privileged over the use of earlier and now cheaper models, regardless of the level of data accuracy required for the research goals at hand (Leonelli 2018a).

These perceptions of what counts as best practice may well not be accurate: they may reflect researchers' prejudices more than actual assessment processes. Either way, they matter enormously to the implementation of OS, with OS tools looking effective only within specific types of research environments and cultures, to the exclusion of others. The preference for specific technologies turns out to depend on factors other than the suitability of that tool to the scientific tasks at hand. Such factors may be infrastructural, such as the availability (or lack thereof) of appropriate training and support for adopting a given technology; institutional, including the structure of scientific publishing and the powerful role played by referees and editors; or sociocultural, like the reputational hierarchies characterizing each field and the common assumption that rich labs should act as role models for other research sites. These factors affect the type of research being conducted, with researchers reluctant to explore potentially useful tools due to the perceived stigma attached to their use. They also inform collaborative strategies, as researchers who do not have access to resources and infrastructures viewed as essential for international publishing often choose to partner with richer institutions who may provide such access – or decide to publish their results only locally, or not at all. Hence the visibility, reputation and even self-consideration of given pieces of research depends on access to high-end technology, with technological preferences embodying specific systems of research assessment, resourcing and geo-political location. While FOSS and cheap sequencing technologies are recognized as valuable in theory, their use in practice clashes with existing – and sometimes conflicting – assumptions about what counts as reliable science, and who gets to decide. This does not make FOSS and related initiatives any less valuable, but rather indicates the importance of taking context into account when considering what systemic changes to the research landscape may be required to support FOSS adoption.

3.3 The Data Trade: Crop Data Linkage and Bioprospecting

Another crucial factor for OS implementation is the political and economic landscape in which research elements are disclosed and traded. Consider the dissemination of research data about crops, and its relation to the circulation of conceptual assumptions about what constitutes sustainable cultivation and the significance of high-yield crop varieties. Crop data, generated in abundance by researchers as well as breeders and farmers around the globe, are crucial to research on planetary health. Bringing data about plant genomes, physiology, growth patterns and environmental responses together can inform novel strategies to ensure food security, including re-imagining agriculture away from high-yield monocultures and using insights into the biodiversity of crops and their resilience to climate change to boost sustainable cultivation and conservation efforts around the world. There has therefore been substantive investment in ways through which plant data may be linked and collectively mined, regardless of where they have been originally collected (Williamson and Leonelli 2022). Given the vast heterogeneity in the sources and materials from which data are extracted, and the difficulties in developing formats and infrastructures that can appropriately document such diversity, data linkage in this area remains an immense challenge (Williamson et al. 2023). When focusing primarily on technical issues around data sharing, however, researchers often fail to address its broader context – including the political economy of data trading across local breeders, national governments and industries with a stake in farming, and the use of evidence from plant science to foster an understanding of agricultural development that is focused on technologically fueled solutions to increase plant yield through genetic selection (e.g. precision agriculture, Miles 2019). Making data widely available on global databases, with little regard for what regimes of intellectual property (and resulting innovations) may apply down the line, carries risks for the farmers and breeders generating data through their labour and expertise. It is often unclear what benefits, if any, the indigenous and farming communities who contribute to data collection may accrue – not just in financial or reputational terms, but in terms of being able to engage in decisions around how data will be used in the future.

These issues have implications for the theoretical assumptions around plant biology that underpin the development and use of crop data infrastructures. For instance, much of the crop science fuelled by open databases privileges the sharing and analysis of data about plant genomes (so-called digital sequencing information) to identify crop varieties that display resistance to pathogens or environmental stressors, and ensures that those varieties are commercially developed and traded. This exercise is not epistemically nor ethically neutral. By favouring the circulation of decontextualized genetic sequences and related plant materials (germplasm), this approach systematically devalues information about the local provenance of such objects, including their environmental and socio-economic context. Genomic data are given priority over observations made on plant phenotypes and uses within local settings, and even the most wide-ranging data collections are standardized to foster smooth comparison among locations, often at the expense of cultural, environmental and biological differences. This does not mean that the plant knowledge of breeders and local farmers is disregarded entirely: rather, such knowledge is appropriated, organized and rendered through the lens of the priorities and taxonomies utilized by plant researchers – and especially molecular approaches. Despite valiant effort to broker fairer forms of collaboration between data producers and users in this domain (for instance by the Consultative Group on International Agricultural Research Communities of Practice, whose understanding of OS I discuss in Section 5), which include a re-imagining of how plants can be studied and understood (Leonelli 2022b), the governance of crop data access and reuse remains by and large under the control of a restricted group of data experts, working with a specific understanding of agricultural development predicated on the identification and cultivation of high-yield crops. This results in a skewed scientific understanding of crop biology and ecology. Moreover, the commodification of insights acquired through such decontextualized data continues to go unchallenged – upholding the well-established agronomic trend of transforming locally acquired information into expensive products (seeds, fertilizers, pesticides) that are then sold back to farmers at a high price (Bonneuil 2019, Curry 2022). Under these conditions, the use of Open Data becomes yet another form of bioprospecting, namely an exercise in extracting resources from underprivileged locations to the advantage of large corporations working in the agricultural sector (Hayden 2003, Benjaminsen and Svarstad 2021).

Thus on the one hand, the quest for extensive plant data linkage is motivated by the desire to explore the ecological features of plants used for human consumption, and relatedly, different models for agriculture, including the advantages of subsistence agriculture and local uses of legacy crops and agrodiverse cultures. On the other hand, such an exploration is conceptually and practically limited by data systems which systematically favour genetic data over other sources of evidence, support market-led models of food security and fail to address the inequities characterizing crop research carried out in the North and the South of the world, or to confront the long colonial history of exploitation that underpins much of crop science to date (Leonelli 2022a). Data organizations such as the Research Data Alliance and international bodies such as the Food and Agriculture Organization have long recognized this problem and are attempting to disentangle the call for Open Data from the practice of bioprospecting. This requires sophisticated forms of data management – including both technical tools and political governance – that foster data trade for agricultural improvement while acknowledging the rights and perspectives of indigenous groups, farmers and breeders (Williamson and Leonelli 2022). At stake are decisions over what counts as significant data, when, where, and for which purpose – as well as over which models, methods, algorithms and publishing format may best support the analysis of such data. These decisions tend not to be driven by dialogue among the many communities with relevant expertise, but rather by technical and commercial concerns that delimit what is accepted as legitimate plant knowledge. Linkage strategies aimed to improve data sharing can thus unwittingly flatten the epistemic space within which plants are studied and managed, with existing regimes of agricultural development erasing the very diversity – biological, cultural, environmental – that the data are meant to document (Leonelli 2022b).

3.4 Methodological Clashes: The Reproducibility 'Crisis'

My last example concerns the raging debate around the principle of reproducibility. Reproducibility, broadly understood as the ability to replicate existing research in ways that yield consistent results, is often presented as a pillar of

OS in at least two ways: first, it seems to demand the sharing of data, methods and code, without which it is arguably impossible to engage in replication in the first place; and second, it is expected to help discriminate between credible and dubious research results, thereby certifying the quality of published results (Burgelman et al. 2019; National Academies 2018, 2019; Leonelli and Lewandowsky 2023). Over the last decade however, a series of high-profile failures to reproduce seminal studies in psychology and biomedicine raised serious questions around the credibility of published results more generally. These concerns are exacerbated by perceived failings in quality controls exercised by journals, difficulties in ruling out fraudulent or questionable research methods (such as p-hacking and selective reporting) as well as lack of clarity over who may be responsible for checks over the reliability of results published online. In this climate of mistrust, reproducibility is often invoked as criterion to distinguish science from pseudoscience, with non-compliant research being viewed as potentially unreliable (Open Science Collaboration 2015). Highly controlled and standardized experiments with pre-specified goals, such as randomized clinical trials or gene knockout experiments on model organisms, constitute well-recognized instances of reproducible research and are held up as models for good research practice more generally. The data and protocols produced through such controlled settings are also among the easiest to share and reuse, given that they are often obtained in digital form and accompanied by consistent metadata (Leonelli 2018b). But where does this leave research settings where controls are not as strict, and where an excessive degree of standardization may jeopardize investigation altogether, by overdetermining researchers' interactions with their objects and thus their chance to garner novel/surprising insights?

Philosophers have noted how reproducibility takes different forms and meanings depending on which cluster of methods, skills, settings, data types, targets, conceptual assumptions and goals turns out to be of relevance to any particular project (Radder 1996, Romero 2019, Guttinger 2020). Even within the same discipline, there can be dramatic differences in the significance ascribed to replicating a computer simulation, where control over research settings is high, given their artificial nature, and both procedures and results are expected to be fully reproducible; field-based observations, where there is little control over research settings, and what is reproduced is often the observer's skills rather than the results themselves; or experiments conducted under changing (environmental, social, climatic) conditions, where the detection of variation is the starting point for new investigations, rather than an indictment on the methods being used (Leonelli 2018b, Feest 2019). Taking highly controlled experiments as universal models for best practice in reproducible research, against which other forms of research are evaluated and results are demarcated as more or less credible, can therefore be damaging. Similarly problematic is the assumption, often made by proponents of reproducibility as a marker for research quality, that research projects should have precisely defined goals from the outset – a situation exemplified by clinical trials meant to test a well-defined hypothesis, but which does not fit exploratory research aimed at identifying and characterizing phenomena of interest.

This overly narrow interpretation of reproducibility risks to extend the current climate of mistrust to *any* non-scripted judgment made by investigators during research, whether such judgements are grounded in hard-won expertise and understanding of the objects at hand or not. This devalues the role of skilled expertise and embodied knowledge in research production, processing and assessment, as well as the significance of social context. Hence, as we also saw in the cases of OS implementation mentioned in this section, efforts to support a universal view of reproducibility entail a potentially damaging understanding of (the boundaries of) domain expertise. Taking a monolithic understanding of reproducibility as a demarcation strategy for the whole of science sidesteps the precious plurality of methods developed to suit specific goals, concepts and target objects. To a researcher who spots differences between repetitions of the same study, asking 'why is this result different' can sometimes be more valuable than asking 'where is the mistake'. Indeed, appeals to reproducibility alone do not help researchers to distinguish between the many possible explanations for a contested result, which can range from differences in research conditions to unintentional mistakes, intentional cheating or the ingenious refutation of a generally accepted fact. Nor do appeals to reproducibility always help address long-term questions around the reliability and quality of results, since they do not help tackle systemic issues with scientific publication cultures and the lack of credit for data stewardship, which arguably push researchers to overgeneralize and under-check their findings. A narrow conception of reproducibility may look attractive as a simple and general solution to the thorny problem of assessing research quality, but its failure to recognize and value methodological and scientific diversity can severely damage scientific advancement, while also proving unhelpful in addressing systemic cultural and institutional problems such as the scarce rewards for validating results and the inequities permeating OS initiatives in domains like crop science.

4 Rethinking Values: Diversity and Justice Across Systems of Practice

Every example considered in Section 3 presents some respects in which research situations may differ from one another, and which are directly relevant to the content and quality of the knowledge being produced. I shall hereafter refer to such differences as *epistemic diversity*, which I define as *the condition or fact of being different or varied in ways that affect the development, understanding and/or enactment of knowledge.*[13] In this section, I argue that consideration of epistemic diversity, including how it is managed within any one system of research practice, is: (1) vital to understanding how science works; (2) inseparably tied to specific interpretations of epistemic (in)justice; and (3) a starting point for any effort to conceptualize and implement OS, without which the OS aspirations towards research quality and transparency cannot be fulfilled.

4.1 Epistemic Diversity and Systems of Practice

The first step in my argument is to note the variety of elements involved in implementing OS principles and tools, and the different ways in which such elements can be clustered and aligned within specific situations of scientific inquiry. The examples in Section 3 highlight the crucial role played by infrastructural, institutional and sociocultural factors, such as those listed in Table 2, in determining not only the conditions of possibility for research, but also the criteria used to evaluate its procedures and results. This is worth emphasizing since some of these factors are not always regarded as having epistemic import, with some philosophers assuming that conceptual or methodological components of research can be considered and evaluated in isolation from social or material elements such as institutional settings and infrastructure. By contrast, I take the attempts mentioned in Section 3 to implement OS as illustrating the interdependence between elements traditionally considered as having direct scientific import – such as theories and methods – and elements sometimes regarded as 'external' or 'accessory' to research – such as geographical location, intellectual property regimes and administrative support. Following in the footsteps of Helen Longino, Alison Wylie, Heather Douglas and Miriam Solomon, among others, whose work has long probed this issue, I contend that drawing a strict distinction between what may count as internal or external factors is problematic precisely due to the high levels of epistemic diversity characterizing research. Rather, such a distinction needs to be drawn on a case-by-case basis, through a situated understanding of which aspects of a given research environment affect scientific goals, methods and outputs at any stage of an investigation.[14]

Closely related to this argument is the observation that disciplinary boundaries are not the only, nor perhaps even the primary, markers for epistemic diversity

---

[13] Definition adapted from the Cambridge English Dictionary.

[14] See Leonelli (2016) for the significance of defining the 'context' of investigation in relation to specific situations of inquiry – building on research by early pragmatists and feminist philosophers.

Table 2 Sources of epistemic diversity of relevance to OS, classified under six umbrella categories: material, conceptual, methodological, infrastructural, sociocultural and institutional.

CONCEPTUAL
- ▲ Theoretical perspective
- ▲ Background assumptions

MATERIAL
- ▲ Target objects
- ▲ Materials
- ▲ Provenance

METHODOLOGICAL
- ▲ Methods and modelling tools
- ▲ Standards: formats and semantics

INFRASTRACTURAL
- ▲ Funding levels and constraints
- ▲ Information and communication technologies, and other digital technologies
- ▲ Venues for publishing and exchange
- ▲ Mobility and transport
- ▲ Funding sources and related commitments

SOCIOCULTURAL
- ▲ Systems of research assessment (local, national, international)
- ▲ Legal and ethical accountability
- ▲ Geo-political location
- ▲ Values and goals
- ▲ Language
- ▲ Demographic characteristics of researchers (gender, class, ethnicity, age, physical ability)

INSTITUTIONAL
- ▲ Career stage
- ▲ Power dynamics
- ▲ Institutional and administrative support
- ▲ Field of study and related norms
- ▲ Intellectual property regimes

within scientific research. There is no doubt that disciplines remain indispensable units of knowledge making, with a crucial role to play both in mandating and in justifying the use of specific clusters of theoretical, institutional and methodological preferences (Mäki et al. 2018). Yet references to disciplinary traditions do not suffice to capture the capillary, highly situated nature of epistemic diversity, as evidenced by the local variation encountered within the same disciplinary spaces, where researchers working under different conditions and on different questions may have widely different perceptions of what constitutes best practice and who is responsible for adjudicating it (Leonelli 2012, Gerson 2013, Levin et al. 2016). Moreover, appeals to disciplinary training and location do not always capture the diverse ways in which researchers may organize their work to confront a given situation of inquiry (Andersen 2016, Nersessian 2022). This is especially notable given the global increase in interdisciplinary and transdisciplinary research efforts aiming to confront systemic challenges such as climate change, pandemics and population growth. Within such multi-perspectival projects, understanding the role played by epistemic diversity within OS and its implementation involves a finer-grained analysis of differences among research approaches than that offered by the broad categories of 'discipline', 'domain' or even 'field'. To this aim, I deploy Hasok Chang's idea of *system of practice*, which denotes any 'coherent set of epistemic activities performed with a view to achieve certain aims' (2013, 16).[15] This framing of scientific activities focuses on the performance of research at any given moment, and the ways in which inquiry is structured and carried out to pursue specific goals – which can be narrow or broad, static or changeable, widely shared or privileged by few. The breadth and flexibility of this approach facilitates a granular analysis of 'the condition or fact of being different or varied' and of what makes such differences epistemically salient ('affecting the development and/or understanding of knowledge' as in my definition of epistemic diversity), beyond assumed and institutionalized categories such as 'discipline'. Building on current philosophical scholarship on scientific pluralism, I now discuss four characteristics of systems of practice which can help conceptualize and implement OS in ways that valorize, rather than undermine, epistemic diversity.

Systems of Practice Are Specific to Local Conditions, Goals and Targets

First, as evident from Chang's definition, systems of practice differ in their *specificity* to local conditions, targets and goals. At their best and following iterative refinement over time, research strategies and tools are exquisitely tailored to suit the

---

[15] Barnes provided a definition for the term 'practice', which usefully complements Chang's analysis and my own usage of the term: "collective accomplishments of individuals concerned all the time to retain coordination and alignment with each other to bring them about" (2001, 33)

characteristics of the phenomena under investigation. Hence methods, theories and models differ depending on their suitability to target objects (Mitchell 2003) and the availability of materials exemplifying that target (Wylie 2003). For instance, a paleontologist who does not have access to remains from the Cradle of Humankind in South Africa, one of the richest sites for humanin fossils from the Plio-Pleistocene, will not be able to investigate ancestors of *Homo sapiens* such as *Australopithecus sediba*; and once she gains access to such materials, she will most likely revise her existing understanding of human ancestry to suit new findings. As novel humanin remains continue to emerge from sites in southern and eastern Africa, and technologies used to analyze those remains proliferate accordingly, the questions asked by researchers will change, and so will the skills and expertise required to investigate them, along with the very phenomena considered to be central targets of inquiry (Currie 2018). Similar situations, whereby researchers discover or change their targets amid an ongoing investigation, characterize research agendas and strategies in most domains, particularly those centered on the study of living entities which develop, evolve and socialize in ways that are often unpredictable. To best study our dynamic, processual world (Dupré and Leonelli 2022), systems of practice thus need to be problem-oriented (Love 2008), adaptive (McLeod and Nersessian 2013) and responsive to the inherent instability of targets (Feest 2019, Massimi 2022). These requirements stand in tension with the OS tendency to assume stable targets and foreseeable goals for research, and foster generalizable approaches and standard methods as desirable constituents of best practice. As effective as standardization is in enabling the sharing and comparison of insights and resources, it needs to be calibrated against the value of system-specific features of local research settings, whose specialized features makes them most sensitive to changes in the parts of the world under investigation.

Systems of Practice Have Different Degrees of Entrenchment within Repertoires

Over time, some systems of practice acquire a reputation for being more reliable, easier to mobilize and/or more productive than others. This can be due to their effectiveness in achieving goals, the robustness of their methods, their suitability to existing policies and institutional settings, and/or other factors. Such systems of practice, including the clustering of values, beliefs, institutions, methods and goals associated with the study of the phenomena in question, may thereby become entrenched as 'gold standards' for research concerning those phenomena (Caporael et al. 2013). In some cases, this results in a system of practice becoming institutionalized as a research field in and of itself (Hackett et al. 2017). In others, systems of practice become what Ankeny and Leonelli (2016) call *repertoires*:

ways of doing science that do not necessarily align with disciplinary boundaries but retain a strong influence as blueprints that can be easily and widely adopted and are implicitly recognized as effective and reliable.[16] As we saw in Section 3, a case in point is the use of randomized controlled trials as an exemplar for how reproducibility should be conceptualized and assessed; the way in which data-intensive crop science is set up to serve precision agriculture, with its reliance on genomic sequencing technologies and its commitment to identifying high-yield varieties, is another.

In principle, it could be argued that good research practice involves the freedom to consider which system of practice may be best suited to investigating given questions and targets (whether such a system already exists or needs to be developed from scratch). In practice, such freedom hardly ever exists: there are strong incentives to redeploy existing repertoires, not least because such mature systems of practice tend to have a standardized structure – including well-developed OS infrastructures – and require less work than the creation of a new system. Repertoires thus often come to define scientific 'success' and canalize understanding of 'best practice' – as for instance in the perception of proprietary software or high-throughput sequencing as proxies for the reliability and quality of the research at hand. This in turn explains the observation that the more a researcher achieves, the more visibility she is bound to receive – a phenomenon which Merton dubbed the 'Matthew effect' and defined as 'the accruing of greater increments of recognition for particular contributions to scientists of considerable repute and the withholding of such recognition from scientists who have not yet made their mark' (1968, 57). Merton emphasized the relation between the Matthew effect and the limited number of individuals who are institutionally recognized as top scientists – where, as (notoriously) in the case of Nobel prizes, the fortunes of a selected few tend to rise exponentially while equally worthy candidates are left in relative obscurity.[17] Attention to repertoires highlights a complementary explanation for the effect: whenever someone's approach becomes recognized as an exemplar of best practice, that recognition tends to result in increasing power and resources (for instance in the forms of awards and funding), which then further strengthens the hold of that system of practice as a repertoire for others to adopt and perform.[18]

---

[16] Repertoires may comprise elements as disparate as skills, concepts, instruments, materials, strategies, structures required to enact projects. What matters is not the co-existence of these elements within the repertoire, but rather their role in scaffolding researchers' performance – the way they think and act. I come back to this point in the next section.

[17] Collins' (1998) monumental study of intellectual change similarly stressed 'limited attention space' as a reason for some individuals gaining more traction than others.

[18] See also Intemann (2020) and what Ross-Hellauer et al. (2022) call 'cumulative advantage'.

The extent to which standards for making or evaluating research are embedded in a wider repertoire is highly relevant to OS. Repertoires provide a significant scaffold for some systems of practice, and this may result in the entrenchment of aspects of the repertoire in the very definition and understanding of what counts as research in that domain, with significant epistemic import for which methods, goals and expertise receive support and acclaim. Indeed, the second characteristic I want to underscore is that systems of practice differ in the degree to which they are *entrenched within existing repertoires*, and thus the degree to which researchers are free to select and develop systems of practice that are specific to their target objects.

Systems of Practice Differ in Their Permeability to Newcomers

The specificity and entrenchment of systems of practice, when considered together, present a problem for OS. The standardization and redeployment of existing resources, including data and software, is a priority for OS – particularly when interpreting openness as sharing, thereby prioritizing transparency and free access as a fundamental step towards improving research practice and communication. However, the quest for standardization and redeployment is also a key avenue by which systems of practice lose specificity and epistemic diversity. Researchers working with a system that is entrenched within existing repertoires may not value – or even consider – elements that are not already part of that repertoire. And even when wanting to modify a repertoire, researchers may face significant hurdles – in the shape of negative reviews, rejection by funding bodies and critical questioning by powerful peers. Of course, focusing critique on new proposals is often warranted, for the sake of validating new methods and corroborating new ideas; such scrutiny is at the heart of science as a safeguard against dogma and groundless speculation. However, the modification of established repertoires is not always a matter of radical innovation, but rather of acknowledging ways of doing research which, while already tried and tested, have not yet gained widespread recognition. As we saw in Section 3, for instance, the study of agrodiversity, including consideration of the long-term ecological implications of growing specific plants variants in particular locations, has long been recognized as a crucial component of crop science, and yet has not played a leading role in structuring infrastructures for data linkage in that domain. While considerable scientific effort is now directed towards modifying existing data systems to incorporate information about environmental effects, this work is hampered by the prominence of genetic information as a central kernel and legitimizing force for current systems of practice and related views on agricultural development.

This speaks to the considerable normative thrust exercised by research repertoires on the everyday conduct of research. As Joseph Rouse pointed out in his seminal discussion of research systems, 'a practice is not a regularity underlying its constituent components, but a pattern of interaction among them that expresses their mutual accountability' (Rouse 2002, 48). In other words, all systems of research practice encourage and stabilize a specific kind of normativity, which in the case of repertoires becomes the basis for communication and collaboration among participants over an extended period of time. This in turn involves specific strategies to manage novelty and adjudicate which modifications do or do not fit with the existing system, which is crucial to the effectiveness and integrity of the repertoire's distinctive way of doing. Here is the third characteristic of research that I wish to highlight: systems of practice differ in how they define and manage their *permeability to epistemically relevant newcomers* (whether these be ideas, methods, people, technologies or research sites), with conservative approaches presenting a distinctive challenge to the openness and inclusivity promoted by OS policies.

Systems of Practice Are Grounded on Specific Demarcation Strategies

This brings me to my fourth and final point, which concerns the *demarcation strategies* used within any one system of practice to determine whether results can be reliably regarded as scientific contributions, and who should be involved in such decisions. Whether such demarcation strategies are implicitly assumed or explicitly discussed, their development and adoption by researchers is an unavoidable part of creating and maintaining a system of practice in the first place. Systems of research practice are systems of demarcation and exclusion. By setting criteria for what constitutes proper science and what does not, and which forms of expertise are deemed to be relevant, demarcation strategies provide the glue that brings and keeps epistemic activities together – what makes systems of practice coherent, in Chang's terminology, and keeps repertoires stable, in Ankeny's and Leonelli's.[19] This was famously recognized by both Karl Popper and Thomas Kuhn, though Kuhn's notion of paradigms, with its insistence on large-scale, incommensurable change, failed to capture the fine-grained, situated and dynamic nature of demarcation; and Popper dismissed the normative relevance of factors other than the conceptual and methodological, grounding his demarcation between science and pseudoscience on the universal mechanism of falsification – an unhappy choice given that many or even most research efforts

---

[19] Note that while convergence on a common agenda is often an important component of demarcation strategies, it may not be necessary since other factors (e.g. agreement on specific methods of evaluation) can also ground demarcation and ensure the coherence of epistemic activities.

are not attempts to corroborate or falsify a bold hypothesis, and what constitutes a falsification remains hopelessly underspecified (Hesse 1974, Lakatos 1978).

In contrast to Kuhn's and Popper's take on the problem of demarcation, my discussion of systems of practice is intended to highlight the epistemic significance of *situated* decisions around what should or should not be part of a system of practice, *given specific conditions and goals*. Indiscriminate appeals to general principles or procedures for 'best practice', including those underpinning the OS movement, do not map easily onto research strategies and systems on the ground. For a system of practice to produce reliable knowledge over time, the ability to adapt to researchers' changing environments, understanding and motivations is crucial, as is the ability to evaluate and compare the choices made by other systems of practice with similar goals and/or set-up (Gerson 2013). The iterative revision of a given system's demarcation strategy is therefore part and parcel of good scientific practice. And while generalizations and standards make it easier to compare systems of practice and even to collate systems together as part of the same broader repertoire, they also tend to obfuscate system-specific differences which may turn out to be salient to future investigation.[20] As Richard Levins (1966) remarked with respect to biological models, there is an inescapable trade-off between generality, realism and accuracy, which needs to be carefully monitored particularly given the epistemic premium typically placed on generalizable tools, methods and claims.

Today's scientific landscape is striving for ways to connect, integrate and perhaps even unify what looks like a hopelessly fragmented and hyperspecialized knowledge base, thereby supporting a model of discovery as a collective effort to accumulate and integrate insights. Open Science policies promise to enhance collaboration in ways that support such efforts. This quest is defensible only if an awareness is retained of the epistemic costs and losses involved, and mechanisms are in place to critically evaluate, on a regular basis, the exclusionary logics underpinning whatever criteria are being used to assess what does or does not belong within the system at hand, and thus what is or is not sanctioned as an acceptable way to do research. Hence debates around OS implementation within any one system of practice need to include explicit and regular consideration of existing demarcation strategies – who and what is included and why, what criteria are being used to make judgements about relevance, whether such criteria have been updated to reflect the latest scientific and social developments, and what the possible consequences of applying such criteria may be in the longer term. Failure to carry out such assessment can have dire epistemic

---

[20] I am not claiming that hyperspecialization is making it impossible or undesirable to devise and implement general standards; rather, that the unavoidable tensions between such standards and situated practice need to be regularly examined, sometimes resulting in revisions at both ends.

implications, particularly for fields where both social and scientific assumptions around a given phenomenon are changing fast in response to new insights and shifting cultural perceptions. Consider for instance clinical research that is committed to a rigid use of binary gender categories (such as women/men) as biological variables, a situation that clashes with non-binary understandings of gender and is arguably hampering research on gender differences (Nature 2018). Or think again about the tensions underpinning different approaches to sharing plant data to foster agricultural development. These examples highlight the profound and fast-shifting epistemic diversity (ranging from empirical methods to social contexts) characterizing systems of practice that operate within the same domain. They also illustrate the normative weight attached to deliberations around demarcation within each of those systems, to which I now turn.

4.2 Epistemic Justice as a Stepping Stone for OS

When a system of practice becomes entrenched and widely adopted as a trustworthy repertoire, it is all too easy for the exclusionary logics presupposed by that system to be black boxed and accepted as 'best practice', with no investment in understanding how they may affect future research. Pluralist and feminist philosophers have long pointed to the dangers posed by this form of conservatism to the reliability of knowledge claims. As Sara Ahmed argues, 'Use can lessen the plasticity of function: when spaces become more comfortable by being repeatedly used by some, they can also become less receptive to others' (2019, 44). In other words, the more something (a tool, a viewpoint, an authoritative reference, a way of doing research) is deployed, the less space may remain for alternatives. This general tendency can have pernicious consequences within the sciences, where we would expect contributions to be evaluated for their epistemic merits and relevance to the investigation at hand, rather than the frequency with which they are used or – as exemplified by the Matthew effect – the visibility already accrued by those who champion them.

It is widely acknowledged that the start of inquiry necessitates a clear focus, with investigators needing to make decisions around how to initiate a scientific project and on which targets. There is no a priori reason why initial choices should fully determine which demarcation criteria are used in later stages of research. And yet, it has been repeatedly observed that lack of representation for any one perspective in the first stages of an investigation leads to its systematic neglect in subsequent research, *irrespective of its epistemic value* – a phenomenon that Philip Kitcher suggestively called 'nonrepresentational racket' (2001, 129). This conservative tendency is also a key target for Helen Longino, whose attention to the condition under which results are scrutinized is grounded on a normative commitment to 'transformative criticism' – that is the will to revisit and, when appropriate, change evaluative criteria. In her words, 'not only must potentially dissenting voices not be discounted; they must be cultivated' (2001, 132). Exclusions based on social conventions embedded in successful repertoires, such as the perception of open research software as less likely to be favourably reviewed, can be particularly damaging since they may not have a scientific rationale and yet they have powerful epistemic implications. Arguably, OS practices should combat this tendency by explicitly challenging the dominance of long-standing repertoires, regularly verifying the value of the components of existing repertoires, and actively encouraging inclusivity where relevant and warranted, even where this complicates attempts to develop common standards and infrastructures.

This brings me to the issue of epistemic injustice, which I consider to be as significant to the conceptualization and implementation of OS as the issue of epistemic diversity, which I discussed in the previous section. Standpoint theorists have pointed out that research, particularly as performed within Eurocentric scientific institutions, tends to exclude the knowledge held by certain social groups in favour of already dominant perspectives; that the groups excluded from science tend to be the same groups marginalized by Western society, including women and queer scholars, people of colour, and political dissidents; and that such exclusions can dramatically reduce epistemic diversity and, with it, the chance of considering and confronting different perspectives on the same phenomena (Harding 2015, Massimi 2022). The type of injustice at work here, which stems from – and has consequences for – research processes aimed at generating knowledge, is what Miranda Fricker famously characterized as epistemic injustice: 'wrong done to someone specifically in their capacity as a knower' (Fricker 2007).

The exclusion of farmers and breeders from the management and analysis of crop data is an example of how assumptions about who constitutes a reliable knower can affect a system of practice. In that case, many parts of crop science regard breeders as outsiders, whose knowledge and expertise are excluded from consideration as professional research activities. Fricker refers to such a situation, where there is a strong prejudice against taking someone seriously as a research contributor, as a case of 'testimonial' injustice; Massimi (2022) calls this an instance of 'epistemic severing', to emphasize the extent to which such prejudice prevents any consideration of the targeted expertise as a potential source of insight for scientific research. Another, complementary form of epistemic injustice is what Fricker calls 'hermeneutical', namely the marginalization of specific ways of thinking and knowing to the point that they are perceived as unintelligible and misguided.

Hermeneutical and testimonial expertise often go together. A case in point is the skepticism towards qualitative research often displayed within debates on reproducibility, which sometimes portray qualitative methods as hopelessly subjective and devoid of rigorous forms of data collection and verification. This assumption is *hermeneutically unjust* towards the centuries of hard-won, sophisticated methodological expertise cultivated within those fields, including the methods devised to probe the external and the internal validity of a given inference; and it is *testimonially unjust* towards researchers who utilize such qualitative methods, who are sometimes regarded as second-class scientists – when they are accorded the status of scientists at all. Another example, also discussed in Section 3, is the assumption that all biomedical researchers involved in sequencing pathogen variants have the same capability to engage in digitalized data sharing and reuse, and therefore to benefit – directly or indirectly – from making their data open. This assumption, which data sharing initiatives such as GISAID have attempted to challenge, unjustly obliterates the needs and perspectives of researchers working in low-resourced environments, whose capacity to contribute to international research may be hampered by Open Data systems set up to work with high-end technologies and analytic tools. This in turn diminishes the comprehensiveness and representativeness of data collected at a global level as well as the range of theoretical perspectives brought to the analysis of such data – a clear case of injustice generating a significant scientific loss.

These examples highlight the extent to which epistemic injustice is interwoven – and typically, inversely correlated – with epistemic diversity. Assuming that qualitative research traditions in anthropology or sociology are incapable of rigorous research means drastically reducing the diversity of systems regarded as exemplars of good practice. Similarly, assuming that breeders cannot be reliable contributors to research on crops means excluding their perspective and expertise from crop science, which is problematic given its relevance to addressing the questions posed within that field. The more a system of practice is prepared to reconsider its own boundaries and demarcation strategies, thereby reassessing the extent to which it can incorporate diverse sources and viewpoints, the more that system will mitigate epistemic injustice, which in turn enhances the system's ability to generate novel, reliable knowledge. This is not to say that systems of practice should be constantly questioning their own assumptions and participants, which would quickly bring research to a standstill.[21] Rather, such questioning can and should happen at regular intervals to match relevant developments in research and society. As the questions

---

[21] I also don't mean to imply that systems of practice should respond to any line of critique, no matter how outlandish or justified; although even in highly instrumentalized cases such as climate change and vaccination, some level of engagement with critics is arguably generative – as long as there is a genuine attempt at communication and reciprocal understanding. Since

being asked, the knowledge being held, and the phenomena being analysed change, so may the forms and sources of expertise relevant to investigation. The very ability to compare and triangulate different sources of evidence, which is widely recognized as fundamental to knowledge production, is grounded on consideration of demarcation strategies: who/what is included in the conversation, when and how; what is accepted as reliable evidence; what crosses borders, and why (Harding 2015, Oreskes 2019, Cartwright et al. 2022). Paying attention to the relation between epistemic diversity and epistemic justice is thus crucial to the reliability and robustness of research results.

As I already remarked, Popper was right to identify demarcation as a key challenge and defining condition for science as a practice and as an institution. In ways that strongly resonate with today's concerns around misinformation, Popper recognized the difficulties of eliminating 'pseudoscientific' elements encroaching on research practice, while at the same time keeping science participative and non-dogmatic. Crucially for my purposes, he also recognized the interdependence between ideas and practices of openness in science and in society, and particularly the political and philosophical challenge of devising governance that fosters individual scrutiny and freedom of expression, while also enabling consensus and progress. Popper's answer to these challenges was to pursue context-independent forms of demarcation, facilitated by what he called 'piecemeal social engineering' designed to instigate critical debate in an incremental, modular manner (Popper 1945). Far from being a top-down system of research governance, this would be a loosely related collection of institutions with responsibility for overseeing and supporting research and its role in society, each of which would need to be revised and constantly adapted to the changing reality of both science and society. Such piecemeal engineering is compatible with, but not dependent on, democratic rule. While democracies provide a space within which scientific freedom can be negotiated and ratified in relation to broader societal requirements, Popper envisaged scientific institutions as needing to be multiple, diversified in their approach and motivations, and at least partly independent from the vagaries of representative politics.

Popper's influence on science policy within advanced liberal democracies, and particularly on how several European governments and the European Commission decided to champion OS, can hardly be underestimated. Many European initiatives to foster OS are explicitly geared towards a federated, distributed approach whereby many different small and medium-sized

---

I cannot engage with the ongoing debate on public engagement and 'science wars' here, I refer to De Melo-Martín and Intemann (2018) and, in relation to OS, Elliott and Resnik (2019).

initiatives are linked and coordinated, without necessarily being subsumed to one another. This system goes a long way towards fulfilling Popper's vision of piecemeal social engineering, including in its partial disconnection from democratic politics. Popper saw a degree of autonomy as necessary for scientific institutions to foster the kind of collective, critical, constructive scrutiny required to achieve reliable knowledge, and thus – he thought – preserve their commitment to pursuing truth above and beyond ideology and partisan politics. The question for contemporary OS policies is whether this vision of context-independent science has been taken too far, resulting in a dangerously idealized view of scientists' priorities and background. Defending scientific research from instrumentalization by vested interests seems more relevant than ever, at a time when understandings of the impact of humans on the planet are highly polarized and politicized. However, it may be argued that some OS policies instantiate a vision of the autonomy of research as requiring isolation from social values, and particularly consideration of diversity and justice, no matter how those may be relevant to the content and pursuit of scientific knowledge. For instance, while there is much talk of affirmative action and of promoting the work of vulnerable groups vis-à-vis the long-standing dominance of white patriarchy among leading academics, the OS system supported by European institutions tends to focus mostly on institutional diversity and technocratic solutions. The emphasis is first and foremost on developing and promoting tools and infrastructures, such as international consortia, interoperable infrastructures and standards to make data findable and reusable. The European Open Science Cloud (EOSC), a highly ambitious effort to coordinate access to European research data infrastructures, is a good illustration of such trends. The EOSC uses a federated model to foster interoperability among disparate data initiatives, thereby protecting existing domain-specific databases and reflecting the disparate interests and goals of their existing funders. The overarching goal is to make research data as easily accessible as possible, fostering efforts to mine data across domains and locations at a scale hitherto unthinkable. At the same time, the scale and institutional focus of the initiative leaves relatively little space for efforts to support epistemic diversity and epistemic justice within everyday scientific practice. The hope is that such benefits will accrue once a functional, effective and well-governed data infrastructure is in place, which begs the question of what forms of injustice may be plaguing the research communities involved in those efforts.

This is a logical consequence of understanding openness first and foremost as an invitation to share resources. As I have illustrated through reference to concrete examples of OS implementation, and existing philosophical discussions on the role of diversity and demarcation strategies within research, the conceptualization of openness as sharing fails to recognize existing inequities – including various instances of epistemic injustice – within systems of practice, and does not help to address the pernicious effects that elitist and conservative forms of inquiry exemplified by dominant repertoires and their demarcation strategies are having on the quality of scientific outputs. One way to address this concern is to place efforts to identify and mitigate epistemic injustice at the centre of science governance and related OS efforts, thus highlighting them as a necessary starting point for implementing OS, rather than the hoped-for conclusion of a journey that starts with indiscriminate sharing and relies on institutional governance to appropriately shape how knowledge is extracted from the resources made available.

The pursuit of truth requires discrimination, and so does the practice of openness. Researchers are constantly making hard choices – around which objects to study, which instruments and methods to trust, which bodies of knowledge to consult, which goals to aspire to. Among those choices are decisions around what to make open, to whom, when and for which reasons. To date, some parts of the OS movement – particularly its institutionalized, top-down incarnations – have paid too much attention to designing procedures and technologies for sharing, and this has come at the expense of strategies, training and procedures to assess who is included and excluded from such apparatus, understand why and with which implications, and mitigate eventual instances of epistemic injustice. To correct this trend, I propose to invert this conceptualization of the direction of travel for OS implementation, which, as I argued in Section 1 (see Figure 1), sees efforts to make research transparent as the starting point to improve its quality and, eventually, its capacity to include. Instead, the implementation of OS needs to start from consideration of what it may take to make research more inclusive, diverse and just – rather than expecting such an outcome to naturally follow from the 'right' choice of software, infrastructures, standards, publishing platforms, or whatever other technological or institutional fix is being devised to facilitate access to resources (Figure 2). It is only through explicit consideration of the demarcation strategies presupposed and supported by OS systems that research quality can be reliably evaluated, and transparency pursued in ways that are informative, discerning and suited to the research context in question.[22] In the

---

[22] Onora O'Neill, and later C. Thi Nguyen, argue for an unavoidable trade-off between transparency and trust, whereby the demand for transparency tends to trigger deception and eventually loss of trust among the people involved (O'Neill 2002), thus engendering a surveillance regime that admits no contextuality and subjectivity (Nguyen 2021). I instead view the relation between trust and transparency as context dependent. A strategic approach to transparency, whereby some resources are disclosed while others are kept under wraps, can go a long way towards increasing trust. This however depends on the overall credibility of the institutions/groups involved in making those choices, and the efforts taken to justify them to multiple publics. A morally

Figure 2 Core values in OS implementation: the proposed direction of travel (which inverts the direction illustrated in Figure 1).

next section, I consider what this inversion of priorities means for the very idea of openness and its role within research, before turning to concrete instances of such a philosophy of OS in my conclusion in Section 6.

5 Rethinking the Philosophy of OS

Having argued that the conception of openness as sharing is flawed, my next task is outlining an alternative conception of openness grounded on considerations of inclusion rather than transparency. This is the task of this section. To this aim, I first need to investigate the philosophical roots of the OS emphasis on sharing and transparency. In what follows, I argue that it involves a problematic conceptualization of scientific inquiry as the effort to appropriate and mobilize outputs, thereby making them valuable. In other words, the understanding of openness as sharing is predicated on an object-oriented view of science, where the availability of commodified, stable, tradeable resources is what determines how researchers use those objects to obtain new knowledge. By contrast, I propose a philosophy of openness predicated on a process-oriented view, whereby research is understood first and foremost as an effort to foster collective agency, grounded on intimate forms of relationality and trust, among widely diverse individuals and groups – an agency that is often enacted through recourse to various technologies, shared interpretations of research outputs and collaborations with non-human agents.[23] This view of research, grounded in social epistemology and the empirical study of scientific practices, understands openness as the quest for *judicious connections* among researchers – connections that are always mediated by the exchange of

---

bankrupt government can make sensible decisions around which data to release, and to whom, and still be widely mistrusted, while a reputable corporation may choose to inappropriately publish sensitive data, and yet benefit from trust acquired on other grounds, especially when time has been taken to explicitly justify such decisions.

[23] While I recognize that non-human entities, including organisms and machines, have their own agency which often contributes to shaping research practice, I here focus on the interactions among humans tasked with developing and evaluating the goals and outputs of science and technology.

objects and technologies but can never be subsumed to such an exchange, lest science loses the power to support meaningful human interactions with an ever-changing world. Such a philosophy of OS, I contend, is both a better description of what it takes to conduct research and a better normative stance on how knowledge should be generated.

5.1 Openness as Sharing and the Object-Oriented View of Science

We have seen how one may understand the idea of sharing underpinning many OS efforts as the opportunity to gain unlimited access to resources that are considered relevant to scientific investigation. Such resources may include certain ways of doing research, such as methods, techniques and skills, but are most often understood to involve objects: models, codes, data, samples, publications. These objects are precious, it is assumed, because they constitute the prime materials and tools from which knowledge can be extracted: their value lies in their potential to inform future research. Data, in particular, are treasured as prospective sources of evidence, with the expectation – so apparent in today's obsession with the revolutionary power of Big Data – that the more data are generated and made available to researchers, the higher the chance that those data can be interpreted and transformed into well-corroborated knowledge claims (Leonelli 2016).

This understanding of knowledge production, which I shall call the object-oriented view of science, is tied to the Western predilection for inductive reasoning as a crucial source of empirical insight. From Aristotle's fondness of observations to Francis Bacon's invitation to collect and analyse 'brute facts', the history of Western scientific efforts – and related institutions – has been largely grounded on the appropriation and manipulation of research outputs, whereby discovery is construed as a more or less linear progression from the gathering of facts, texts, measurements, observations, materials to the generation of new knowledge claims. Of course, this is a general trend with many exceptions, as evidenced by the significance of theories and various forms of deductive reasoning in many areas of research. Yet I think it fair to acknowledge how inductive inference, broadly construed as the extraction of insight from systematic consideration of available objects (typically construed as 'sources' which are systematically collected by powerful research institutions), has long been favoured as fundamental to the very idea of empirical knowledge; and to note how this object-oriented understanding of how science may progress encourages the compulsive accumulation of more and more resources from which knowledge can be extracted, thereby feeding into what we now understand as a capitalist model of human development predicated on constant growth and speculation over prospective profit (de Sousa Santos and Meneses 2020). Christophe Bonneuil (2019) pointed to this mode of inquiry as a 'resourcist' approach to understanding nature, whereby the very notion of biodiversity is construed as the collection and study of genetic resources, that is organisms whose genetic materials can be isolated, mobilized and transformed to suit human goals (as in the case of crop data linkage).[24] Similar assumptions underpin the collection of many other types of materials, methods and data: for instance, satellite images as critical documents of the evolution of planetary health, whose continuous accumulation and analysis can inform – and transform – both the natural and social sciences; or the ensemble of modelling methods for tracking and predicting the spread of infectious disease for human and non-human organisms, whose comparison may yield an overarching understanding of pathogenic threats as they move and develop.

Within this view of scientific inquiry, access to existing scholarship becomes a condition of possibility for any investigation, which makes the activity of sharing into an obvious focal point for OS implementation. It is not within the purview of this Element to provide a systematic critique of the object-oriented view of science, nor do I wish to deny its significance and tremendous success in eliciting new ways to make sense of the world and intervene in it. What I wish to emphasize are two of its main characteristics, which turn out to have severe implications for understandings of openness, scientific practice and research governance. The first is a general *distrust of human cognitive abilities*, and particularly of the role of history – both the individual biographies of researchers and their social and institutional context – in shaping human understandings of the world.[25] Bacon famously warned against the Idols of the Mind, including the ways in which reasoning may be affected by culture, social status and language, as symptoms of the human propensity to distort and shape one's experience of the world to suit one's circumstances and hopes, thus diminishing the value of the resulting knowledge. The very idea of subjectivity was thereby construed as a separation between subject and objects of research, whose study presupposes a divestment of interests and values in favour of a neutral 'view from nowhere'. Over the last three centuries, the wish to take human bias out of science has increasingly taken the form of efforts to automate discovery, most recently through reliance on artificial intelligence (AI) tools geared towards minimizing human error – making space for what Lorraine Daston and Peter Galison (1992) called mechanical objectivity. It is no

---

[24] Among the extensive literature supporting this view, see Helen Curry on maize cultivation (2022) and Hannah Landecker's (forthcoming) on the use of patents to harness and appropriate enzymatic metabolism for mass-scale industrial production.

[25] I am assuming an enactivist, embedded, embodied and extended model of cognition.

coincidence that one of the most prized characteristics of research components shared online is machine readability: the expectation is that the easier it is for AI systems to hoover up and process data, the better trained such systems will be to recognize and assess diverse inputs, resulting in more sophisticated analysis. Hence a plant recognition system trained on imaging data extracted from 10,000 plant species is expected be more accurate and reliable than a system trained on data from 1,000 species; and an algorithm trained on demographic data at the national level is expected to yield better results than one trained on data from one city.

More generally, the object-oriented view of science is often associated with a belief in the power of research methods to rescue human judgement from bias and cognitive failings, thereby providing ways to validate inductive inferences as objective and context independent. The promise of big data springs from the expectation that research results, once adequately cleaned up, processed and standardized, can be safely taken as accurate representations of the world, whose validity and evidential value can be assessed regardless of the circumstances under which data are produced and used.[26] The examples that we considered in Section 3, however, cast doubt on this expectation. What they demonstrate is that the standards used to assess data quality and meaning are shaped by domains and social contexts characterized by unequal and sometimes unjust relations among those involved in their production – a fact that needs to be considered when using such standards to assess diverse research situations. Whether a plant recognition system is reliable, and for which purposes, may depend on what is assumed to count as a valuable trait as much as it depends on the volume of imaging data available; and appropriately curated demographic data from a small territory, where adequate steps have been taken to ensure the representativeness and accountability of the sample and data processing tools, may be preferable to a much bigger – but uneven – dataset from a larger region. Hence considerations of epistemic diversity and justice need to underpin the interpretation of research components. Access to such components is no guarantee of appropriate reuse. On the contrary, the further a research object travels from its context of origin, the more difficult it becomes to assess whether and how the demarcation strategies underpinning its production and processing can serve the new situations (goals, settings, participants) within which the object is being deployed. Thus, data standardization does not necessarily support fully automated data analysis over trained human assessment: just as making data machine-readable fosters wide dissemination and uptake within AI systems, standardized formatting can heighten the need for researchers to exercise skilled

---

[26] Leonelli (2016, 2020) provides a detailed critique of this view.

judgement, on a case-by-case basis, to evaluate the adequacy of data as evidence base for novel purposes.

The second characteristic of the object-oriented view of science that strongly affects current understandings of openness is the centrality of the idea of *ownership*. Just as early modern scientific institutions thrived on the colonial appropriation of objects from around the world, which were collected and stored by Western museums and scholarly societies in the hope of informing scientific investigations,[27] contemporary OS infrastructures collect, manage and distribute objects viewed as relevant to knowledge generation. Ownership does not need to involve long-term or even exclusive possession of these objects. I rather take it as indicating the ability to control some uses of the objects at hand and manipulate their characteristics accordingly, which is taken as an indispensable prerequisite for the practice of research (quite literally in the sense of making them 'one's own'). Debates over who has ownership and control of research outputs – as well as objects that, while not produced during research, are deemed to be useful to research efforts, such as social media data – continue to be central to any effort to conceptualize and implement OS. In some cases, such debates take the form of requests to relinquish ownership claims. For instance, publicly funded researchers may be asked to donate their models, methods and data to online infrastructures with no expectations of return or recognition, in the name of collaboration and transparency, while groups who are not directly involved in knowledge creation, such as farmers or medical patients, may be asked to donate methods, materials and data for the benefit of society as a whole, as encapsulated by the motto 'sharing is caring' (or, somewhat perversely, the 'right to science'). In other cases, the debate focuses on ways to assert ownership through agreement on specific conditions for exchange, which may consist of legal protections such as licencing agreements or technical procedures to govern access (as in the case of GISAID). Either way, debates around intellectual property, as the site in which ownership is exercised and decisions are made about what resources are available for further investigation, constitute the epicentre of both the conceptualization and implementation of openness in many quarters of the OS movement.

That the understanding of openness as sharing involves placing ownership claims at the heart of OS may seem paradoxical, but it is certainly not a new claim and is corroborated by the history of OS practices. In his seminal account of the emergence of open software in the 1970s and 1980s, Christopher Kelty highlights how the initial impetus to escape proprietary forms of software development quickly evolved into an understanding of openness as 'freedom

---

[27] Such as the Royal Society, in an explicit effort to implement Baconian empiricism (Walsh 2018).

to buy' (Kelty 2007, 149–51), thereby identifying intellectual property as key condition as well as 'blind spot' for open systems (ibid., 178). Kelty is not alone in underscoring the parallels between ideas of free exchange, which underpinned the rise of the OS movement, and neo-liberal support for the free market, with its emphasis on information sharing as mechanism for future prosperity and for the acquisition of control over outputs. We also saw this in the 2015 definition of OS by the European Commission, where OS was explicitly presented as an enabler of innovation and economic growth. It is indeed tempting to consider how the OS landscape is endorsing and strengthening what Manuel Castells called 'information capitalism' and Nigel Thrift labelled 'knowledge capitalism' – in other words, a knowledge system grounded on continuing speculation over the projected growth of commodified information and unequal access to technology (also Wyatt et al. 2000). It is not a coincidence that the circulation of data, now widely advertised as the 'new oil' fuelling economic growth, continues to be a fundamental goal for OS initiatives in both the public and the private sectors; and as Philip Mirowski (2018) has argued, participation in OS implementation, ranging from Open Access publishing to Open Methods, can – and often does – increase the profits of large corporate actors such as publishing companies and pharmaceutical industries.

But was OS not supposed to disrupt the commodification – and related trends towards closure and secrecy – of scientific research? And where do OS initiatives explicitly geared towards attacking dominant regimes of ownership over research components feature in this reading of the philosophy of openness? As I discussed in Section 2, the disruption of mechanisms and institutions geared towards the appropriation of research objects continues to be a key motivation for many of the researchers and non-profit organizations involved in the OS movement, whose construal of OS practices and infrastructures is explicitly aimed at bypassing ownership claims.[28] Within such initiatives, the idea of sharing is interpreted as *unlimited reuse*, rather than unlimited access: the emphasis is on disrupting existing constraints on people's capacity to work on and with research components, including constraints as different as border controls, institutional boundaries and intellectual property regimes. In that respect, OS initiatives take inspiration from social movements and activist lobby groups, especially in the extent to which they use various forms of protest to challenge existing power dynamics, engage in sustained interactions with

---

[28] This impetus often motivates appeals to the FAIR principles, which aim to make data Findable, Accessible, Interoperable and Reuseable. Despite endorsement by several policy and research organizations, FAIR remain difficult to implement, not least since implementation requires questioning who owns the data, who finances the required infrastructures and who assesses whether specific instances of reuse are responsible and viable.

institutions and develop an alternative vision for the future – all of which eventually turns them into significant political actors themselves (Della Porta and Diani 1999, Leonelli 2019b). In this view, any attempt towards appropriating research components may be regarded as problematic, and indeed many of these initiatives over the last three decades have been characterized by a refusal to engage with ownership claims altogether – with the idea that progress in research is obtained when bypassing existing property structures and thereby breaking conventions around who owns what, and for which purpose, in favour of a blanket permission to (re)use anything that may serve the process of discovery. Again, scientific attempts to support the linkage of crop data serve as a useful example, since many prominent initiatives to share such data do not directly engage with national and corporate regimes of ownership over research outputs and related materials (such as seeds and germplasm), but rather ignore such issues in favour of a focus on how best to include participants and share resources (Leonelli 2022a). The COVID-19 platform for genomic data exchange has a similar focus on making data accessible and actionable to as many as possible, with little regard for which forms of appropriation and commodification may be tied to such sharing efforts – and how this may affect data donors' perception of sharing infrastructures as trustworthy and fair. This attitude is reflected in the broader history of biological data sharing. The global steer towards free and immediate dissemination of sequencing data, enshrined in the so-called Bermuda Rules, famously emerged from ownership disputes around the results of genome projects carried out in the 1990s, whose principal goal was avoiding data privatization (Maxson Jones et al. 2018): free data access and reuse were given priority, but this came at the expense of debates over how such sharing fosters patenting and corporate strategies for biomedicine and agriculture. These concerns were taken out of scientific discourse around Open Data and out of the purview of OS advocates. The result has been the establishment of an epistemic economy for science which is grounded on the possibility of transnational and transdisciplinary data exchange, and yet remains largely oblivious to the regulatory, legal and economic regimes under which such exchanges take place, and their implications once research findings are downstreamed into commodities.

Prima facie, the idea of sharing as unlimited reuse constitutes a much better foundation for OS than the idea of sharing as unlimited access. For many researchers participating in OS efforts, the attempt is to explicitly bypass the capitalization of research and its outputs, and instead frame them as common goods, which should be available to anybody who may need them. The fact that simply providing access is not enough to guarantee the productive deployment of resources is well-recognized in those circles, which is in itself an important improvement over the rhetoric of focusing on access without regard for how specific forms of access may inform or even determine subsequent use. Nevertheless, there are marked similarities in the epistemology of science envisioned by those who view sharing as unlimited access and those who emphasize reuse. Both camps tend to portray scientific knowledge as an ensemble of modular components, which need to be packaged, circulated and assembled in a variety of ways in order to foster discovery and novel insights. While the interpretation of sharing as reuse seems to defy associations between regimes of ownership and appropriation and the pursuit of reliable knowledge, the focus on sharing outputs as commodified objects remains unchallenged, as is the idea that OS is predicated on the choice to retain or relinquish control over scientific components – itself a modular view of research as consisting of the assemblage of (vetted and approved) building blocks. Reuse is premised on access to and manipulation of objects, thus replicating the fundamental epistemological presumptions of the object-oriented view of science. There is also a similar distrust in human cognitive abilities and a fear of bias as manifesting itself at the point in which research components are assembled and interpreted – resulting in an insistence on making such moments of assemblage more transparent, with the expectation that this will enable checks and thus make the analysis more trustworthy. Notably, the critical spotlight is most often placed on publicly funded research, in what Manuela Fernandez Pinto calls an 'asymmetrical treatment' tied to the idea that what is publicly funded should be held to a higher standard (2020, 8). This assumption is deeply questionable, given the well-documented problems with science carried out under commercial secrecy (Oreskes and Conway 2010) and the crucial role played by private actors in developing and distributing goods for mass consumption with no accountability for the long-term, systemic effects of those activities on planetary health (Landecker, forthcoming). More generally, and despite efforts to bypass issues of ownership and capitalization in favour of just making science better, there is a fundamental tension in some parts of the OS movement between the pursuit of transparent, free reuse of research components within publicly funded research and the lack of interest in translational research and the downstream commercialization of open resources. Intellectual property thus continues to play a central role in this interpretation of openness, made even more conspicuous by the absence of intellectual property as a focal concern for the scientists involved. Whether OS activists dwell on this or not, what supports many data sharing initiatives – in the literal sense of guaranteeing financial support by governments and funders, as well as evidence of 'impact' – is the expectation that the research outputs being shared will eventually be commercialized and become sources of profit for those with the resources and power to control existing legal, regulatory and economic regimes.

My analysis has reached a stark conclusion. Whether openness is conceptualized as an effort towards appropriation or disruption, access or reuse, underpinning these seemingly opposing camps is a common vision of the 'sharing' of objectified resources as the starting point for scientific research. This vision underestimates the extent to which the objectification of constituents and outputs of research is a temporary assumption made by researchers to define and demarcate systems of practice for particular purposes, rather than an ontological affirmation of what the world is like. Elsewhere, John Dupré and I have described this process and its potential dangers as a form of *means reification* whose instrumental and situated value is too often forgotten in the rush to use research outputs as trustworthy mirrors of reality (Dupré and Leonelli 2022). The genetic sequences that biologists have collected on coronavirus strains are expected to play a specific epistemic role: to help identify potentially harmful mutants. Whether such data can be reliably interpreted as representing the biology of viruses more generally will depend on the purposes and conditions of future research, as well as researchers' skilled assessment of whether the viral samples from which data were extracted can credibly represent a complex, ever-evolving and ever-diversifying microbial environment. The object-oriented vision of science tends to set worries around the representativeness and long-term reliability of research components aside, underestimating the significance of the judgements that scientists are required to make whenever they decide to reuse such components for novel purposes. Research components are conceptualized as items whose quality and usability can be verified independently of specific circumstances, thus facilitating their immediate deployment towards novel discoveries – a view of research components as commodities that colludes with a framing of scientific epistemology as the extraction, control and accumulation of epistemically valuable objects. I maintain that this view has limited capacity to underpin OS practices in the long run – and in its effort to depict human deliberation as external to good research, it is liable to bypass concerns around epistemic justice and diversity as irrelevant to the pursuit of reliable knowledge.

5.2 Openness as Judicious Connection and the Process-Oriented View of Science

Let us now consider what it takes to conceptualize openness in ways that are less closely tied to the commodification of research outputs and an object-oriented epistemology. We need a different starting point: a process-oriented view of research, within which science is not primarily concerned with owning and controlling objects, but rather with the ability to act skilfully in ways that (1) support the human capacity to understand and interact with the world, and (2) can be communicated, adopted and verified by groups other than the ones responsible for any given discovery.[29]

The process-oriented view of research is well-suited to a conception of knowledge that Chang calls 'active': that is the ability to perform and coordinate certain activities and interventions in ways that are purposeful, even if the aims of such coordinated performance are not well-defined – or are deemed to be liable to change – at the start of inquiry. Most importantly for my purposes, this view of knowledge-as-ability 'should not be reducible or subordinate to the storage and retrieval of information' (Chang 2022). Of course, active knowledge does encompass the use of objects such as data, publications and models to store and codify information. The advancement of science as a collective endeavour would not be possible without the collection and/or development, manipulation and interpretation of objects to represent given parts of reality and/or enable specific interventions in the world. Such objects are the forms through which active knowledge is abstracted, encapsulated and traded – thus acting as essential material anchors to communication and the interpersonal, social nature of scientific inquiry. In this sense, any research practice necessarily relies on texts, graphs, models, observations, measurements and other artefacts through which active knowledge can be exchanged, evaluated and modified; and proponents of OS are right to stress the exchange and sharing of objects as relevant to good scientific practice. However, the production and trade of research objects should not be construed as a primary goal of science. The overarching purpose of sharing these objects is not simply to convey information, but rather to facilitate human agency, whether in the form of reasoning (resulting in scientific explanations or theories) or interventions (producing methods and tools to interact with the world). Sharing data can thus be useless when researchers have no way to assess their evidential value, or unnecessary when investigation focuses on overarching patterns emerging from the data (data models) rather than the characteristics of individual data points. Similarly, making research protocols accessible is not that helpful when lacking the ability to use relevant instruments, training and infrastructure.

Nor should the sharing of research objects be viewed as a way to transform what are quintessentially human artefacts, whose very identification as relevant

---

[29] As in the case of object-oriented epistemology, the idea of a process-oriented epistemology has deep roots that I cannot discuss here (Dupré and Leonelli 2022). My approach builds on the philosophy of science in practice, which conceptualizes research as an ongoing process (Soler et al. 2014).

to inquiry is situated in time and space and liable to be challenged in the future into neutral products that can be interpreted irrespective of their provenance and the ways in which they have been processed. Whether and how objects provide information to investigators depends on the conditions and goals of inquiry, as well as relevant understanding of the history of those objects and of the motivations and backgrounds of those who participated in their development (Morgan 2010). Even material samples such as fossils and biological specimens are framed, stored and processed according to specific expectations around what they may represent and what insights they may foster (Currie 2018, Ankeny and Leonelli 2020, Wylie 2021). These objects, just like digitalized data and mathematical models, are but a material snapshot of a particular moment in research practice: they are not meant to be counted as timeless scientific outputs in and of themselves, but to act as situated scaffolds for epistemic activities aimed to increase active knowledge. Their validity, relevance and significance within research thus need to be routinely reassessed and adjusted to the relentless changes within science, the world and human aspirations. An immediate implication of this approach to scientific knowledge is to concede its positionality vis-à-vis human goals. If we take the overarching aim of research to be the cultivation of skills – including situated ways to frame and explain reality – that can foster human understanding and inform interventions in the world, we cannot treat the generation and use of research objects as free of normative assumptions. Rather, the positionality of contributors and the diversity of perspectives brought into scientific pursuits need to be recognized and explicitly discussed as inextricable from the research process.[30] Hence the process-oriented view places concerns with epistemic injustice and diversity, including a systematic interrogation of the epistemic and social implications of the demarcation strategies underpinning systems of research practice, at the heart of science, rather than relegating such concerns to its periphery.

This framing of scientific epistemology has implications for the conceptualization of openness within research. In what follows, I suggest that a process-oriented view moves the spotlight of OS away from the interpretation of openness as sharing, directing it instead towards an interpretation of openness as the *establishment of judicious connections among researchers* – connections that are typically mediated by and constituted through technology and familiarity with specific research settings, including human and non-human participants in those settings. To unpack this idea, let us start with discussing the notion of connection. The Cambridge English Dictionary provides three main definitions

---

[30] Wylie (2003); see also the work of Code, Harding, Longino, Douglas, Elliott and Massimi, among others.

for this term. The first focuses on relationality, with connection depicted as 'the state of being related to someone or something else'. The second places more emphasis on how the environment within which a relation is established can help to scaffold that relation: connection is then 'the act of joining or being joined to something else, or the part or process that makes this possible'. The third definition highlights the cognitive state of entities who are involved in establishing a connection: 'a feeling that you understand, like, and are interested in someone or something'. All three of these dimensions are essential to my understanding of connection as the core activity and goal of OS. Connection is, first and foremost, a process of engaging another entity or being in ways that may potentially breach one's assumptions and established practices (one's own demarcation strategies, in the case of research). This process of engagement is scaffolded in a variety of ways, ranging from the technical means and concrete objects through which the connection is made to the emotional demands of forging social bonds and opening up to the challenges that novel relations may offer to one's existing understanding of the world. Indeed, the process of connecting is the more challenging, the more different the 'other' being engaged is to oneself. The idea of openness is quintessentially linked to that of learning as going beyond one's boundaries. This arguably applies to systems of research practice as much as to research groups and individual learners: establishing new connections often means expanding one's learning, challenging existing assumptions around what is considered external or irrelevant to a given system, and considering whether new boundaries need to be established that incorporate the novel relations. The environment surrounding any given situation of inquiry, including the many objects, creatures and materials that scaffold the quest for knowledge, provides the overarching landscape within which forming and maintaining a connection becomes possible. And human cognitive and emotional states – including an interest in connecting and exploring what this may mean for one's own sense of identity, social relations and assumptions about the world – are unavoidably part of that landscape.[31]

To give flesh to these abstract ideas, consider two examples of OS practice that do give a central role to the forging of connections. One focuses on supporting short-term, contained connections: this is exemplified by the emergence of platforms such as Crowdfight which aim to help researchers identify someone willing and able to help them with a specific, concrete task, encourage collaboration around that task, and oversee the process of distributing credit. The objective is to build solidarity and attention to diversity through time-limited volunteering, which researchers may be

---

[31] While I cannot discuss this here, I endorse a view of emotions as fundamental to human cognition (see Colombetti 2014).

able to fit in their schedule depending on their working conditions and willingness to expand their connections and accountabilities beyond their immediate location. Crowdfight was created in 2020 to connect researchers working on COVID-19 and managed to attract over 45,000 volunteers in its first year of operations, with several innovations and even scientific papers produced as a result. Its success in establishing new connections is clearly related to the worldwide call for scientists to suspend their normal activities to help confront a global and immediate threat. The focus on circumscribed interactions makes the volunteering effort manageable even to extremely busy researchers; most important for our purposes, the emphasis is resolutely on human-to-human interaction to understand and address the problem at hand, rather than on the provision of standards or tools that can help people resolve the problem by themselves. Hence researchers asking for help are not simply sent to a website or a tool and asked to watch a 'how-to-use' video: they are put in touch with people who have experience of using relevant tools and who are willing to help others navigate them and determine whether and how they can be helpful to their questions. Obviously the immense advancement in digitalized communication and collaborative platforms makes this type of connection possible and productive. At the same time, it is recognized that expanding the user base for these platforms requires the creation of human connections to identify new problems and the ways in which existing tools may – or not – serve new purposes and diverse research cultures.

As a system for micro-collaboration, Crowdfight builds on a long-standing Open Source ethos of well-meaning communication towards achieving common goals, embodied for instance by GitHub, which makes it accessible to different publics, including people who do research but have limited or no programming skills. Whether such a mechanism of connection across expertises can function in other domains and in relation to the overarching emergency created by climate change, it remains to be seen. In any case, systems of micro-collaboration offer a counterpoint to the competition-driven scientific landscape, proposing manageable ways to reimagine one's research as inclusive, service oriented and connected to disparate locations and conditions of research. They recognize that confronting everyday problems is an important step towards providing visibility to a wider variety of research experiences, thereby potentially recalibrating the research system towards a more diverse set of participants. This finding aligns with several existing studies of successful grassroot OS initiatives, such as those documented by the Open and Collaborative Science in Development Network, or OCSDNet – a collaboration among several OS initiatives from several countries, including many in the Global South, which found that 'the ability to participate, to connect, and to co-produce knowledge with others who share common concerns is far more important than simply access to content or resources' (Chen et al. 2019, 2).

Following in this vein, my second example consists of an effort to support long-term connections with a view to strengthening key OS infrastructures. This is the establishment of so-called 'communities of practice' to help develop and maintain ways to share and interpret biological data collected across species, institutions and national boundaries (Louafi et al. 2022). Within the plant science domain, one such community is the Consultative Group for International Agricultural Research Ontologies Community of Practice, which was formally constituted in 2017 as a formalization of long-standing international efforts to bring together transdisciplinary, international expertise relevant to the dissemination of crop data (Arnaud et al. 2020). Membership of the community is open to those wishing to support the development of sustainable, responsible and effective means to structure crop data infrastructures. The focus on computational ontologies reflects a shared interest in the keywords and semantic tools used to classify, order and visualize the data, while also delimiting participation to those who already have some understanding of the use of ontologies to organize and mine complex data. Participants in the community are aware that decisions on which ontologies to use are as controversial as they are consequential – both scientifically and socially. A central concern is the extent to which indigenous or otherwise local classifications of plants should feature in ontologies, and how they should be related to scientific taxonomies; another is the nomenclature used to describe crop traits of interest to global and local markets, including whether the use of specific labels can foster or diminish interest in the commercialization of specific crop varieties, and which types of commercialization may be most beneficial to consumers of the crops in question. The Ontologies Community of Practice set out to develop data sharing tools that capture and communicate expertise and viewpoints from different stakeholders, ranging from data scientists to crop researchers, local breeders, agronomists, policy makers and agrobusinesses. To this aim, the activities of the community have focused on establishing connections among these groups, with the goal to enhance mutual understanding and provide opportunities for exchange, and in recognition of the crucial significance of trustworthy communication channels towards developing reliable systems of research practice around the crops. Hence the community of practice strives to provide regular online venues for the exchange of viewpoints and actively fosters participation by experts who may represent different perspectives, leading to frequent confrontations among these groups.

Engagement in this domain can be very challenging, with stakeholders motivated by fundamentally different and sometimes conflicting priorities and goals (data scientists focusing on technical solutions, for instance, while breeders worry about repercussions of adopting a given nomenclature for local markets (Williamson and Leonelli 2022)). Yet, the operations of the Ontology Community of Practice are notable precisely for their interest in expanding the scope of the debate and modifying OS practices accordingly, while at the same time motivating the (sometimes controversial) choices being made and examining the implications for diverse stakeholders. The coordinators of this community recognize that the extent to which their data sharing tools will endure in time, and be accepted as a reference point for future research, depends at least in part on building forms of engagement and consideration of each other's viewpoint, thereby helping to address emerging concerns and build a minimal degree of trust among participating groups. This is not fast work and may not be the most effective path towards data sharing in the short term; indeed, communities such as this are under constant threat by funders and participants alike – their work often portrayed as a short-lived means to deliver highly standardized, stable ontologies for automated discovery systems. And yet, in contrast to the case of Crowdfight, the epistemic value of a community of practice lies in its ability to support research choices and changes in the long term. The cultivation of connections in these forms – as an ongoing and critical component of OS practices – leads to more resilient and inclusive infrastructures and tools, whose exposure to a variety of viewpoints and data sources can better inform future agricultural interventions and assessments of when and how to foster change. Transparency is thereby achieved through inclusive deliberative processes that cultivate trust and a shared understanding of the circumstances under which findings and procedures may be regarded as reliable.

Note that such shared understanding is not equivalent to consensus among all participants in an OS endeavour, and does not necessarily result in agreement among the parties involved.[32] Rather, the focus is on creating new forms of intimacy (which can sometimes take the form of vehement disagreement and misalignments) between the human actors – as well as the technologies, materials, institutions and non-human participants involved in the relevant systems of practice – brought together by OS efforts. This focus has clear repercussions on how OS tools are built, used and governed in the long term. For instance, researchers engaging in OS by consulting an online database may be encouraged to take some time to familiarize themselves with the ways in which the database is set up and the work conditions of those who have produced the data made available for retrieval, rather than simply use the database as a neutral data source. Such an effort may slow down data reuse but also enable researchers to discover which assumptions underpin the ways in which data were generated

---

[32] I share Solomon's (2001) skepticism about the power of consensus. Even in the rare cases where it can be obtained, it may result in the formation of new repertoires with their own demarcation strategies and exclusive assumptions.

and presented, and whether such assumptions make sense within new contexts of data use. This in turn could inform a better assessment of whether and how the data could be handled and interpreted within the new setting, thus making their work scientifically sounder (Leonelli 2016, Borgman 2016, Mayernik 2017). In such a case, attempting to establish a connection with an online tool, beyond simply appropriating the data-objects visible on the site, makes it possible for researchers to contextualize the data therein and better use them to inform their existing practices.

Or consider the case of researchers who are trying to decide how to share their methods with others, and to what level of detail. Choosing an appropriate format, publishing platform and framing for methodological descriptions is not something that can be easily standardized and automated, since it involves considering who may be expected to take an interest in those methods and for which purposes, and framing their presentation accordingly (including making it harder to appropriate methods for nefarious purposes). Considerations around the means and publics of one's research are familiar to any experienced scientist, and yet are often set aside in the rush to use data-powered AI to shortcut such efforts and enable scientists to access and use research objects in a modular fashion, without interrogating the baggage carried by those objects and its implications for future research. Doing away with the history and context of research objects, focusing on the challenges of sharing them rather than the challenges of interpreting them, is an attractive proposition in a world of increasingly distributed expertise, where personal links among research communities are frequently mediated – even substituted – by communication technologies and digital platforms. And yet, the trustworthiness and reliability of those technologies and platforms ultimately depends on the collective willingness to keep scrutinizing their adequacy for purpose and the extent to which they embrace and support epistemic diversity and justice within an ever-changing scientific and social landscape (Lusk and Elliott 2022). This, in turn, means having to invest some efforts into opening the black box of digital infrastructures (Bowker et al. 2010, Nowotny 2021), identifying not only the choices and assumptions of relevance to one's own investigation but also the epistemic communities from which such choices and assumptions have emerged.

This is where the idea of *judicious* connection comes in. Openness in research does not only require the effort to establish connections; it also requires the attempt to evaluate which connections may be relevant and beneficial to the scientific effort at hand, and therefore to assess the potential implications (negative or positive) of such initiatives. Establishing connections unavoidably involves exercising judgement, which in turn involves creating new divisions and exclusions. What I am advocating here is therefore not merely an exercise in socialization, with the misguided expectation that by putting people in dialogue with each other a positive outcome will surely ensue. Nor is it an invitation to epistemic parity, encouraging a free society where 'all traditions have equal rights and equal access to the centres of power', as famously advocated by Paul Feyerabend (1978, 9). As discussed above, the scientific world is riddled with epistemic injustice, and addressing such injustice does not simply mean encouraging epistemic diversity in all its forms and giving all potential participants in the research process equal opportunities. Rather, it requires making choices among possible visions of who can and should participate in research efforts, whose perspectives can and should be voiced, whose ideas of research could and should be supported (what Longino (2003) called 'tempered' equality). There is always a decision made around who will benefit and who may lose out from specific research initiatives, and the more explicit and reflexive that choice is, the better. Consideration of epistemic diversity is different from augmenting epistemic diversity *tout court* – agency involves taking sides, thereby betraying the model of free society envisaged by Feyerabend. In the Ontologies Community of Practice, participants constantly adjudicate conflicts between crop experts and data scientists. The outcome of such adjudication may satisfy neither group, or lean heavily in favour of one over the other, with implications that will need to be monitored and assessed as they unfold. In the case of Crowdfight, the resolution of conflicts is instrumental to immediate needs on the ground, in the hope that participants will heed the invitation to pay attention to other researchers' circumstances and needs, and find solutions that orchestrate standards of best practice with the actual goals and conditions of the task at hand. In both cases the building and maintenance of connections need to be judicious: they require skilled deliberation, whereby the new opportunities offered by the connection in question are evaluated within the contexts at hand. Indeed, openness can itself be understood as a dynamic and highly situated mode of valuing the research process and its outputs, which encompasses economic as well as scientific, cultural, political, ethical and social considerations (Levin and Leonelli 2017).

This performative understanding of openness does not require the establishment of active collaborations among the parties involved. I have already remarked that connections may lead to collaboration as well as to conflict, with no guarantee that novel forms of engagement will engender agreement or reciprocal understanding.[33] Moreover, the pursuit of judicious connections

---

[33] "A flourishing science requires both the focused, and thus less epistemically diverse, approach of normal science (where the devil may be in the details) and the free, and more diverse, exercise of

need not involve the level of coordinated social agency exemplified in the short term by Crowdfight and in the long term by the Ontologies Community of Practice. All it requires is the effort to accompany the development of any novel research components, infrastructures or communities with an exploration of the system(s) of practice that contributed to generate that entity, and the extent to which the demarcation strategies utilized by those systems may differ from one's own. In other words, making a connection involves the attempt to acquire knowledge through which those involved can meaningfully assess whether and how to utilize the novel element for their own purposes. To come back to the previous example: a researcher consulting a novel database to see whether it holds relevant data for her own project is making a novel connection. For that connection to prove generative for her own research, she needs to understand something about the conditions under which the data have been produced. This in turn requires her to consult metadata and information about who has created the database, for which purposes and how – which she absorbs and interprets based on her own expertise, experiences and goals. As part of this process, she has acquired a degree of intimacy and familiarity with the database and related systems of practice, which helps her to decide whether and how the data can fit her own work, and at what costs to her own demarcation strategies. In some cases, connections such as these may, in time, facilitate collaboration around common goals: the researcher may decide to visit one of the labs that produced the data in the first place, for instance, or to participate in the development of the database. But even when such overt collaboration does not obtain, judicious connections bring a new experience of otherness, and the opportunity to modify one's perspective and perception of the world. It is through the establishment of connections, underpinned by an evaluation of what those connection may mean for one's demarcation strategies, that fruitful disagreements and frictions may come to light, and generate novel reactions and insight (Edwards et al. 2011).

This interpretation of openness sets up a critical space for moving beyond the economic definitions of value embedded in the contemporary scientific landscape and many OS policies, and pays due attention to the ways in which diverse interests and commitments affect research practices. What researchers choose to make open, how and with whom depends on the goals, preferences, constraints and institutional settings of the researchers involved, making it difficult to maintain a clear-cut distinction between public and private spheres, or between the various layers of sociality in which research is embedded. Within every choice to share a research constituent or output, assumptions are being

---

critical reflection (where the presuppositions of normal science may be exposed as unfruitful or harmful doctrines)" (Radder 2019, 228).

made around who may be able to access and reuse that object, and how. Similarly, within every choice to incorporate a new object within one's research, thus establishing a novel connection, assumptions are being made around the conditions under which such an object may be trustworthy and useful. What I am advocating is to read such exchanges as instances of connections that are best suited to the quest for active knowledge when accompanied by the critical scrutiny, by those involved, of the respective systems of practice and the challenges that such a connection may bring to one's own assumptions.

This approach to the epistemology of OS is already instantiated by many initiatives within the OS movement, including the examples mentioned in this section, which focus on understanding and supporting specific communities and situated forms of use of research components (whether they be texts, methods, hardware, models, data, lectures, code), rather than developing such components and making them freely accessible without a clear sense of who may in fact adopt them. It also parallels existing legal scholarship on knowledge commons, which is moving away from the idea of commons as shared objects and instead emphasizes their social character as forms of community management, thereby recognizing that taking account of the specific conditions and dynamics of social relations is essential to the use of resources for knowledge production (Frischmann et al. 2014). The existence of such work indicates that my argument for openness as judicious connection is not a novel idea, nor something completely absent from the OS landscape, which is itself so diverse. What I have attempted to do in this Element is to single out this approach and its philosophical underpinnings, and articulate some of its implications for research and its governance, especially when compared to other ways of framing the concept of openness. To conclude, I shall spell out the ways in which understanding openness as judicious connection can overcome some of the challenges linked to understanding openness as the freedom to share, and sketch what this may mean for the future of OS.

First, the spotlight shifts from the pursuit of unlimited access to research components to the nature of the *relations* between research groups and related systems of practice, and thus to which level of intimacy and reciprocal understanding may be best suited to the circumstances and purpose of any given connection. Having access to resources is not conducive to knowledge production unless the right skills, infrastructures, governance and administrative support are available to foster use, which in turn requires a minimal degree of understanding and trust among the parties involved: specifically, the ability to assess whether and how those connections align with one's own system of practice, in which respects, and what it would take to ensure some degree of compatibility, were it found to be lacking. Ensuring that OS participants have relevant venues and mechanisms of consultation and feedback requires significant, long-term investment beyond the technical realization of specific tools – which make it ever so clear that OS is not a cheap and cheerful form of division of research labour, whereby one may achieve faster results through intelligent deployment of digital technologies, but rather a resource-intensive transformation of research, whereby scientific knowledge production may become more robust and inclusive. Considering citizen science is particularly useful here, since despite the emphasis on forging novel connections characterizing this form of OS, some citizen science projects end up using non-scientists as cheap sources of novel data, with no interest nor investment in involving participants in processes of data governance and interpretation (Strasser et al. 2018, Prainsack 2020). This mirrors the object-oriented, extractive epistemology of science which I critiqued in relation to 'openness as sharing', and may arguably strengthen pre-existing demarcations between scientific and lay forms of expertise, rather than helping to assess whether the experiences of citizens may inform and even guide research.

Second, OS is not construed as quintessentially grounded on digital technologies, but rather as involving the critical and constructive scrutiny of how digital platforms can support existing and future research – including the effective use of multiple media within specific *social* environments. Digital media, no matter how sophisticated, are not sufficient to communicate active knowledge and need to be complemented by analogue initiatives, such as – in the case of biological data sharing – exchange programmes through which researchers can visit each other's laboratories and learn new ways to handle instruments, specimens and experimental spaces. Attention to the social also extends to the role of humans within research systems. The focus should not be solely on the preferences and behaviours of individual scientists, but rather on the interplay between those individuals and the various collectives (from local groups to national institutions and international societies) through which their working life is organized – and the extent to which the juxtaposition of different social configurations, which may or may not intersect/overlap, affect individual agency and judgement. Moreover, and particularly in the case of citizen science and Open Source coding, professional research networks tend to intersect with non-scientific networks including activist and lobby groups. Within an OS geared towards fostering connections, such links would need to be explicitly mapped and recognized for their potential contributions to research efforts or – as evident in politically charged debates over vaccination and climate change denial – for their harmful effects.

This brings me to the third characteristic of OS within the 'openness as judicious connection' view: OS interventions are not envisaged as globally beneficial in their repercussions. As any other social transformation, OS is understood as necessarily *divisive*. It should not be surprising to observe that the quest for consensus over what may constitute 'best research practice' is often met with some form of resistance. Any shift in practice is likely to have implications that are good for some participants in the research landscape, and bad for others. Recognizing the exclusionary power of OS initiatives – for instance, by noting that building new connections often involves letting go of existing ones, if only because of the limits in human attention space and the demarcation strategies underpinning any system of practice – is essential for confronting existing forms of inequity and discrimination in research. It involves accepting that value judgements are unavoidable when developing open research and infrastructures, no matter how inclusive the relevant technologies promise to be; and explicitly researching, ideally in collaboration with people with relevant expertise, the advantages and disadvantages of any initiative, no matter how well-intentioned.[34] In other words, OS is not only about the making of connections but also about taking controversial, value-laden decisions around where to go next – and being prepared to revise those decisions in light of failure or unforeseen negative implications.

Following from this point, OS initiatives sensitive to the importance of judicious connections are tailored to the plurality of epistemically diverse systems of practice, in ways that can help cultivate epistemic justice as appropriate to the situations of inquiry at hand. The development of common standards and technological platforms is helpful insofar as it supports localized agency: OS needs to aim for *situated* solutions, such as tools explicitly devised for modification in response to local environments (e.g. Crowdfight or GitHub) and mechanisms for OS participants to provide feedback and participate in OS governance and future development (e.g. communities of practice). Such solutions unavoidably involve a ranking of priorities around who should and can benefit from specific forms of OS, how and with what implications for the broader research landscape. Clearly identifying which users are privileged by any one OS initiative, and providing an explicit rationale for such choices, is a more honest and fruitful way to present OS than acting under the pretence that it is 'good for everybody'. Not only does it enhance the trustworthiness of those initiatives, but it also fosters ongoing assessment of the epistemic positioning and value of such initiatives, and related proposals for change.

This means that choosing who should benefit from OS initiatives involves taking a normative, and often moral, stance – and doing so in recognition of the

---

[34] Far from a new insight, this is the central message of Responsible Research and Innovation, and science studies research on digital transformations.

deep interrelation between epistemic and normative concerns within research practice. Broad appeals to equality, as often witnessed when insisting on sharing resources freely and widely, can be unhelpful since even when resources are accessible to all in principle, not all will have an equal chance to utilize them meaningfully for their own purposes. I therefore support moving away from a distributive idea of fairness in OS and instead fostering an *equitable* approach geared towards mitigating existing inequities and actively fostering the capacity for meaningful uptake among the most vulnerable among the prospective participants. Open Science initiatives have a responsibility to assess whether the characteristics of the research landscape they mean to target foster or impede the uptake of OS, and adapt their proposals to the situation at hand. The recent increase in OS initiatives specifically targeting researchers working in low bandwidth contexts and/or fragile institutions is encouraging in this respect, as is the establishment of internet platforms mindful of environmental and social concerns around the energy and hardware required to run them.[35]

Given my analysis of these key characteristics, I conclude that framing openness as judicious connection is helpful and perhaps even necessary to achieve scientifically and socially beneficial forms of OS, thereby improving existing understandings of 'best practice' in research overall – particularly when compared to the framing of openness as freedom to share. Focusing on connections and the significance of judgement places human decision-making and social contexts at the heart of scientific knowledge production, and particularly of strategies to communicate, collaborate and implement research insights. Without this recentring of OS around its human participants, OS risks becoming

Table 3 Synoptic comparison of the two interpretations of openness I have discussed in this Element.

Openness as sharing	Openness as judicious connection
Unlimited	Relational
Digital	Social
Good	Divisive
Global	Situated
Equal	Equitable
Focused on itemized outputs (objects that can be shared)	Focused on social agency (ways of doing and being with others)

---

[35] See dedicated working groups within the Research Data Alliance and 'Low-Bandwidth Design', KM4Dev Wiki (last modified 18 Feb. 2012), http://wiki.km4dev.org/Low-Bandwidth_Design.

yet another form of techno-administrative control over research outputs and their use as commodities. The framing of OS around the idea of judicious connection serves as a normative foundation for a philosophy of OS, whose full-fledged shape and implications cannot be comprehensively discussed within the scope of this Element, but whose future development can credibly help address the research troubles associated with closed science.

6 Conclusion

I have argued that a key challenge for OS is to productively manage the clash between different interpretations and operationalizations of openness, which emerge from diverse systems of practice with unequal levels of influence and visibility – an inequity which, when it is due to social circumstance rather than to the merit and fit-for-purpose of the systems at hand, generates epistemic injustice and weakens the quality of scientific results. Prima facie, this challenge may appear to be purely practical: an issue of implementation rather than conceptualization of OS. It may also look like a purely ethical concern, with little bearing on the quality and content of scientific knowledge. Both these impressions are wrong. What I demonstrated instead is that the difficulties encountered in implementing OS across diverse research environments are tied to philosophical assumptions about how science does – and ought to – work.

In my view many OS efforts, and particularly institutionalized, top-down approaches, are grounded in an object-oriented view of research, within which openness is understood as the freedom to share – and, in the most sophisticated versions, to reuse – itemized research outputs such as data, models and articles. This approach to the philosophy of OS is not adequate nor desirable. It assumes that increasing the accessibility of outputs will help improve the quality of scientific knowledge and the inclusivity of research practices; and makes the sharing of research components into an aim of science in and of itself, thereby focusing OS efforts on the trade and management of objects. By contrast, I proposed that OS practices can better support the quality of scientific outputs when they focus on the specific ways in which accessibility is provided, and particularly the strategies used within specific research situations to decide who counts as a contributor, how objects should be handled and interpreted, and what goals should be pursued. This framework takes research outputs such as data, models and articles as temporary signposts of the ongoing process of inquiry, whose function is to adequately support communication and learning within and beyond the research community. This is a process-oriented philosophy of science, which calls attention to the conditions under which outputs are produced, disseminated, stored and deployed, and conceptualizes scientific research as primarily aimed to advance active knowledge. Far from being solely a question of sharing resources, openness is thereby conceptualized as the opportunity to make and maintain connections among relevant stakeholders in the research process – whether these be professional researchers, other publics, non-human organisms or machines – in ways that help to develop ever more relevant forms of interaction with the world. Who should count as a relevant stakeholder can only be established in relation to specific research contexts, through judicious discrimination that takes account of the diversity of perspectives of potential relevance to the goals at hand, while at the same time seeking to mitigate the forms of epistemic injustice that may affect research conditions.

Framing OS as a platform for the cultivation of judicious connections brings the focus of OS initiatives on researchers' ways of knowing, doing and being with others. The spotlight shifts to epistemic activities that facilitate critical scrutiny of research components and results, including of the demarcation criteria used by researchers to identify research outputs and adjudicate who constitutes a relevant beneficiary of and/or contributor to scientific inquiry. Encouraging researchers to be more explicit in the priorities set within their systems of practice is critical to improving the transparency and quality of research in ways that are responsive to the scientific and social environments within which research is carried out. Far from constituting an obstacle to the implementation of OS, consideration of both diversity and injustice becomes an essential step towards realizing the aspirations of the OS movement, while at the same time providing an opportunity for OS to avoid capture by dominant research repertoires and defy inequitable, conservative, discriminatory and flawed approaches to research.

How this vision of openness can be effectively realized remains itself an open question. It places heavy demands on researchers – and particularly those working within highly overdetermined settings in which criteria and standards for 'best practice' are well-defined and rarely challenged – to seek connections that may constructively challenge their demarcation strategies and help tailor them to the questions at hand. This is not just a case of finding better technologies to communicate results. It often means challenging existing perceptions of who the publics and participants in science may be, which in turn helps to determine which research outputs to disseminate, when and how. Advocates of OS sometimes point to such demands as part of a fundamental change in research culture, which should be driven by researchers themselves. Yet we have seen how scientific practice is constrained and scaffolded by institutionalized systems of incentives and rewards, through which research is supported and assessed, not to speak of the sizable costs of inclusive OS initiatives – both to set up venues and channels for communication and to maintain those over time – which cannot be shouldered by individual projects or research groups. Implementing OS is not only up to individuals or a matter of developing the right technologies: it is a systemic shift demanding appropriate forms of governance, infrastructure, funding and collective agency. This brings us back to the key questions posed by Popper's reflections on the Open Society: to what extent is the organization of science overdetermined by social conditions? And how can scientific institutions confront the need to demarcate what legitimately belongs to systems of research practice at any one time, without at the same time jeopardizing the porous nature of such systems and the non-dogmatic, participative nature of scientific knowledge production?

One key challenge for OS governance is *scale*. Open Science projects targeted to specific communities, goals and domains can build on existing connections to facilitate novel encounters. A case in point are instruction manuals, databases and field books created to encourage reuse of methods relevant to the study of specific organisms or diseases, since many such initiatives involve a relatively consistent (though geographically distributed) epistemic community where people know each other and have some established way to assess each other's work. Scaling up such efforts involves finding enough common ground to prompt judicious connections, without at the same alienating different systems of practice of potential relevance (Chen et al. 2019). Prospective participants need to assert their autonomy as contributors, while also learning from the initiatives they join – which often means investing into venues and information systems where standards can be contextualized, scrutinized and modified in relation to local goals (Kelty 2019).

Many OS initiatives have answered the challenge of scale by proposing novel ways to formalize and describe scientific labour – ranging from Data Management Plans to pre-registration procedures that capture the reasoning and assumptions underpinning a given research design. This brings another key challenge, that of *bureaucracy*. Such tools can be helpful in contextualizing a specific set of outputs and assessing their validity, significance and future potential, thereby increasing accountability. But pushing researchers to adopt such tools (as typically required when interpreting openness as sharing) increases the administrative and managerial aspects of research, creating an additional layer of paperwork and taking time away from actual investigation. This is well justified when proportional to the goals and circumstances of research, but highly problematic when reporting guidelines are out of sync with research practice. Moreover, when tools like pre-registration are used to check whether scientists have done what they initially promised to do, OS threatens to become yet another way for institutions and funders to exercise control over research practice – a form of surveillance that can hamper researchers' creativity and does not necessarily result in better quality checks, since it is unclear who has the expertise, time and motivation to evaluate these new types of outputs. The view of openness as judicious connection moves away from the use of OS tools to control research and instead focuses on helping researchers to identify and question their own control strategies over the systems they use and investigate – for instance, when using tools such as Data Management Plans and pre-registration to track how research has moved on and why, with no expectation that researchers should stick to a pre-defined script. This in turn demands an extensive reorganization of priorities and evaluative systems for research, which places the development of infrastructures and transdisciplinary dialogue at the centre of the academic ethos, with competition playing a secondary role.

Beyond an internal re-orientation of academic priorities and institutions, the biggest challenge – the elephant in the room – is the extent to which OS efforts are prone to *instrumentalization* by the political forces and economic structures within which science is unavoidably positioned. What is the point of considering whether OS is geared towards sharing or towards judicious connections, when most research outputs are buried in an object-oriented research system where trading and appropriating knowledge is the endgame of any scientific investigation? Does it make any sense to consider how publicly funded research is governed, when its results are eventually appropriated by corporate structures with a set hierarchy of beneficiaries?

There is no underestimating how profoundly the global political economy overdetermines the processes and outcomes of scientific research, whether publicly or privately funded. And yet, I do not see this as a reason to give up on OS – and science itself – altogether. Consider again the case of crop data sharing. It is true that the careful system of data governance developed by communities of practice may be thwarted by aggrotech companies looking to profit from such OS initiatives, placing crop knowledge at the service of socially and environmentally unsound forms of agriculture. It is also true that community-led OS efforts have brought international attention to the exploitative nature of crop data sharing, with well-recognized institutions such as the Food and Agriculture Organization endorsing concerns around the vast capitalization of data within neo-liberal markets. As a result, national agricultural strategies and international agreements such as the Convention for Biological Diversity are placing debates around data licensing, data silos and benefit-sharing agreements on their agenda, thereby underscoring the importance of making OS more responsible and responsive vis-à-vis its participants.

Or consider ongoing debates around author-pays models of Open Access. These models make authors responsible for covering the costs of producing and distributing Open Access publications. When researchers can access relevant funding, this is easy to implement: publishers get paid for their services, authors manage to publish and the results are accessible without paywalls. However, funding is seldom easy to come by and, even when it is, takes resources away from other parts of the research system. Incentivizing the author-pays model is thus likely to exasperate existing vulnerabilities and divides among prospective authors, with disastrous consequences for the overall research landscape. And yet, this is precisely what seems to have happened over the last decade, with prominent OS initiatives such as Plan S seemingly endorsing author-pays models of Open Access in ways that preserve the profit margins of commercial publishers, while not taking account the vast inequity in the financial resources available to researchers across disciplines and locations. However, this situation has not gone unchallenged. Many of the ambassadors of Plan S (including myself) have rallied against the commitment to author-pays models within the scheme, while much has been done to document the cumulative advantage this system confers on researchers based in rich institutions (Ross-Hellauer et al. 2022). As a result, Plan S is actively exploring alternative forms of Open Access publishing, while scholarly societies and institutions are paying more attention than ever to the scientific and social challenges posed by publishing structures and the role of commercial publishers within the research system.

Finally, consider coronavirus research. During the pandemic, questions of data access, accuracy and use became a matter of daily dispute on mass media and social networks, resulting in data infrastructures like GISAID playing an unanticipated sociopolitical role. As discussed in Section 3, this has created frictions within the scientific community, but is also fostering a sophisticated debate over which commitments and goals OS infrastructures should serve, why and for whose benefit (Johnson et al. 2022). In other words, these frictions have uncovered the politics of OS – a situation that has taken some researchers by surprise, causing discomfort among scientists who thought they were creating purely technical, sharing platforms only to discover that such a platform could not exist without a normative vision for the role of science in society. Yet acknowledging the significance of such normative visions is a crucial step forward for research, and one that is starting to emerge in some of the most recent OS initiatives and policies. Organizations like the Research Data Alliance, whose focus a decade ago was to provide technical means for data sharing, have expanded their mission towards fostering scientific engagement among multiple publics, brought together by the recognition that they have expertise to bring to data collection, handling and interpretation. There have been extensive calls for 'intelligent openness' (Boulton et al. 2012, Bilder et al. 2020) focusing less on sharing objects and more on the conditions for such sharing to be responsible. And most recently, the UNESCO (2021) Recommendation on Open Science built on an extensive consultation (to which I participated together with colleagues from Committee of Data for Science and Technology and hundreds of OS organizations around the world) to emphasize the importance of processes over products, collaboration over competition, and inclusion over speed as necessary starting points.

These features of the OS landscape provide hope that conceptual interventions in this arena, especially at this time of large-scale transformation in the ways research is institutionalized, assessed and governed, are worthwhile. The very existence of the OS movement is premised on the recognition that the manner in which researchers communicate and collaborate, and the extent to which they can access and use different tools at various stages of the research process, are matters of central importance to knowledge development. Whether or not OS discourse will endure beyond the current hype, newly created infrastructures and standards will leave a lasting mark on how science is pursued in the future and what roles it plays in society. Long-standing appeals to collaboration and sharing are being reconfigured by the push towards speed and digital automation which has come to define the era of Big Data and AI – and the political economy of research and data exchange characterizing our politically fractured world. In this sense, OS signposts a political, economic and cultural moment with long-term implications for how research is carried out and how it is institutionalized. Open Science initiatives are attempting to alter not only scientific methods and communication models but the very meaning of research and the nature of its outcomes. This underscores the significance of identifying and evaluating the conceptual assumptions made within OS, and the ways in which the performance of openness in research practice can be made scientifically as well as socially and ethically robust.

References

Ahmed, S. (2019) *What's the Use?* Durham, NC: Duke University Press.

Allison, D. B. et al. (2016) Reproducibility: A Tragedy of Errors. *Nature* 530: 27–30.

Andersen, H. (2016) Collaboration, Interdisciplinarity, and the Epistemology of Contemporary Science. *Studies in History and Philosophy of Science* 56: 1–10.

Ankeny, R. A. and Leonelli, S. (2016) Repertoires: A Post-Kuhnian Perspective on Scientific Change and Collaborative Research. *Studies in History and Philosophy of Science* 60: 18–28.

Arnaud, E. et al. (2020) The Ontologies Community of Practice: An Initiative by the CGIAR Platform for Big Data in Agriculture. *Patterns* 1: 100105. https://www.cell.com/patterns/pdf/S2666-3899(20)30139-2.pdf.

Barnes, B. (2001) Practice as Collective Action. In Schatzki, T. R. et al. (eds.), *The Practice Turn in Contemporary Theory*. London: Routledge, 17–28.

Bartling, S. and Friesike, S. (eds.) (2014) *Opening Science: The Evolving Guide on How the Internet is Changing Research, Collaboration and Scholarly Publishing*. Dordrecht: Springer.

Beaulieu, A. and Leonelli, S. (2021) *Data and Society: A Critical Introduction*. London: SAGE.

Benjaminsen, T. A. and Svarstad, H. (2021) *Political Ecology: A Critical Engagement with Global Environmental Issues*. London: Palgrave Macmillan.

Bezuidenhout, L. et al. (2017) Beyond the Digital Divide: Towards a Situated Approach to Open Data. *Science and Public Policy* 44(4): 464–75.

Bilder, G., Lin, J. and Neylon, C. (2020) *The Principles of Open Scholarly Infrastructure*. https://doi.org/10.24343/C34W2H.

Bonneuil, C. (2019) Seeing Nature as a Universal 'Store of Genes': How Biological Diversity Became 'Genetic Resources', 1890–1940. *Studies in the History and Philosophy of Biological and Biomedical Science* 75: 1–14.

Borgman, C. L. (2016) *Big Data, Little Data, No Data*. Cambridge, MA: MIT Press.

Boulton, G. et al. (2012) *Science as an Open Enterprise*. The Royal Society Science Policy Centre report 02/12. London: The Royal Society.

Bowker, G. et al. (2010) Towards Information Infrastructure Studies: Ways of Knowing in a Networked Environment. In Hunsinger, J. et al. (eds.), *International Handbook of Internet Research*. Dordrecht: Springer, 97–117.

Burgelman, J.-C. (2021) Politics and Open Science: How the European Open Science Cloud Became a Reality. *Data Intelligence* 3(1): 1–19.

Burgelman, J.-C. et al. (2019) Open Science, Open Data, and Open Scholarship: European Policies to Make Science Fit for the Twenty-First Century. *Frontiers in Big Data* 2: 43.

Caporael, L. R., Griesemer, J. R. and Wimsatt, W. C. (eds.) (2013). *Developing Scaffolds in Evolution, Culture, and Cognition*. Cambridge, MA: MIT Press.

Cartwright, N. et al. (2022) *The Tangle of Science*. Oxford: Oxford University Press.

Chang, H. (2013) *Water Is Not H2O: Evidence, Realism and Pluralism*. Springer.

Chang, H. (2022) *Realism for Realistic People*. Cambridge: Cambridge University Press.

Chen, L., Okune, A., Hillyer, B., Albornoz, D. and Posada, A. (eds.) (2019) *Contextualising Openness: Situating Open Science*. Ottawa: University of Ottawa Press.

Chubin, D. E. (1985) Open Science and Closed Science: Trade-offs in a Democracy. *Science, Technology, & Human Values*, 10(2): 73–80. https://doi.org/10.1177/016224398501000211.

Code, L. (2006) *The Politics of Epistemic Location*. Oxford: Oxford University Press.

Collins, R. (1998) *The Sociology of Philosophies: A Global Theory of Intellectual Change*. Cambridge, MA: Belknap Press.

Colombetti, G. (2014) *The Feeling Body: Affective Science Meets the Enactive Mind*. Cambridge, MA: MIT Press.

Currie, A. (2018) *Rock, Bone and Ruin*. Cambridge, MA: MIT Press.

Curry, H. A. (2022) *Endangered Maize: Industrial Agriculture and the Crisis of Extinction*. Oakland: University of California Press.

Daston, L. and Galison, P. (1992) The Image of Objectivity. *Representations* 40: 81–128.

Della Porta, D. and Diani, M. (1999) *Social Movements: An Introduction*. Oxford: Blackwell Publishers.

De Melo-Martín, I. and Intemann, K. (2018) *The Fight against Doubt: How to Bridge the Gap between Scientists and the Public*. New York: Oxford University Press.

De Sousa Santos, B. and Meneses, M. (2020) *Knowledges Born in the Struggle: Constructing the Epistemologies of the Global South*. London: Routledge.

Dupré, J. and Leonelli, S. (2022) Process Epistemology in the COVID Era: Rethinking the Research Process to Avoid Dangerous Forms of Reification.

European Journal for the Philosophy of Science 12:20. https://doi.org/10.1007/s13194-022-00450-4.

Edwards, M. A. and Roy, S. (2017) Academic Research in the 21st Century: Maintaining Scientific Integrity in a Climate of Perverse Incentives and Hypercompetition. *Environmental Engineering Science* 34(1). https://doi.org/10.1089/ees.2016.0223.

Edwards, P. (2010) *The Vast Machine*. Cambridge, MA: MIT Press.

Edwards, P. N., Mayernik, M. S. and Batcheller, A. (2011) Science Friction: Data, Metadata, and Collaboration in the Interdisciplinary Sciences. *Social Studies of Science* 41(5): 667–90.

Elliott, K. (2020) A Taxonomy of Transparency in Science. *Canadian Journal of Philosophy* 53(3): 342–55.

Elliott, K. and Resnik, D. B. (2019) Making Open Science Work for Science and Society. *Environmental Health Perspectives* 127: 075002. https://doi.org/10.1289/EHP4808.

European Commission (2016) *Open Innovation, Open Science, Open to the World: A Vision for Europe.* https://digital-strategy.ec.europa.eu/en/library/open-innovation-open-science-open-world.

European Commission (2018) *Open Science: Altmetrics and Rewards.* Final Report for the Mutual Learning Exercise Open Science: Altmetrics and Rewards of the European Commission. https://rio.jrc.ec.europa.eu/en/policy-support-facility/mle-open-science-altmetrics-and-rewards.

Fecher, B. and Friesike, S. (2014) Open Science: One Term, Five Schools of Thought. In Bartling, S. and Friesike, S. (eds.), *Opening Science: The Evolving Guide on How the Internet is Changing Research, Collaboration and Scholarly Publishing.* Dordrecht: Springer, 17–47.

Feest, U. (2019) Why Replication Is Overrated. *Philosophy of Science* 86(5): 895–905.

Fernandez Pinto, M. (2020) Open Science for Private Interests? How the Logic of Open Science Contributes to the Commercialisation of Research. *Frontiers in Research Metrics and Analysis* 5: 588331. https://doi.org/10.3389/frma.2020.588331.

Feyerabend, P. (1978) *Science in a Free Society.* London: Verso.

Fricker, M. (2007) *Epistemic Injustice: Power and the Ethics of Knowledge.* Oxford: Oxford University Press.

Frischmann, B. M., Madison, M. J. and Strandburg, K. J. (eds.) (2014) *Governing Knowledge Commons.* Oxford: Oxford University Press.

Gerson, E. M. (2013) Integration of Specialties: An Institutional and Organizational View. *Studies in History and Philosophy of Biological and Biomedical Sciences* 44(4a): 515–24.

Guttinger, S. (2020) The Limits of Replicability. *European Journal for Philosophy of Science* 10: 10. https://doi.org/10.1007/s13194-019-0269-1.

Hackett, E. J. et al. (2017) The Social and Epistemic Organization of Scientific Work. In Felt, U. et al. (eds.), *The Handbook of Science and Technology Studies*. 4th ed. Cambridge, MA: MIT Press, 733–64.

Harding, S. (1995) "Strong Objectivity": A Response to the New Objectivity Question. *Synthese* 104(3): 331–49.

Harding, S. (2011) *The Postcolonial Science and Technology Studies Reader*. Durham, NC: Duke University Press.

Harding, S. (2015) *Objectivity and Diversity: Another Logic of Scientific Research*. Chicago: Chicago University Press.

Hayden, C. (2003) *When Nature Goes Public*. Princeton, NJ: Princeton University Press.

Hecker, S. et al. (2018) *Citizen Science: Innovations in Open Science, Society and Policy*. London: UCL Press.

Heesen, R. and Bright, L. K. (2019) Is Peer Review a Good Idea? *British Journal for the Philosophy of Science* 72(3): 635–63.

Herzog, L. (2023) *Democratic Knowledge*. Oxford: Oxford University Press.

Hesse, M. (1974) *The Structure of Scientific Inference*. London: Macmillan.

Intemann, K. (2009) Why Diversity Matters: Understanding and Applying the Diversity Component of the National Science Foundation's Broader Impacts Criterion. *Social Epistemology* 23(3–4): 249–66.

Intemann, K. (2020) Understanding the Problem of 'Hype': Exaggeration, Values, and Trust in Science. *Canadian Journal of Philosophy* 52(3): 279–94.

Jensen, J. L. (2020) Digital Feudalism. In Jensen, J. L. (ed.), *The Medieval Internet: Power, Politics and Participation in the Digital Age*. Bingley: Emerald Publishing, 95–109.

Johnson, R. et al. (2022) *Intelligent Open Science: A Case Study of Viral Genomic Data Sharing during the COVID-19 Pandemic*. Report prepared on behalf of the UK Government Department of Business, Energy and Industrial Strategy. Research Consulting. https://assets.publishing.service.gov.uk/government/uploads/system/uploads/attachment_data/file/1118628/intelligent-open-science.pdf.

Kelty, C. (2007) *Two Bits: The Cultural Significance of Open Software*. Oakland: University of California Press.

Kelty, C. (2019) *The Participant: A Century of Participation in Four Stories*. Chicago: Chicago University Press.

Kitcher, P. (2001) *Science, Truth and Democracy*. Oxford: Oxford University Press.

Knorr Cetina, K. (1999) *Epistemic Cultures: How Science Makes Knowledge*. Cambridge, MA: Harvard University Press.

Krige, J. (ed.) (2022) *Knowledge Flows in a Global Age: A Transnational Approach.* Chicago: Chicago University Press.

Lakatos, I. (1978) *The Methodology of Scientific Research Programmes.* Cambridge: Cambridge University Press.

Landecker, H. (forthcoming). *American Metabolism.* Cambridge, MA: Harvard University Press.

Leonelli, S. (2012) When Humans Are the Exception: Cross-Species Databases at the Interface of Clinical and Biological Research. *Social Studies of Science* 42(2): 214–36.

Leonelli, S. (2016) *Data-Centric Biology: A Philosophical Study.* Chicago: Chicago University Press.

Leonelli, S. (2018a) Global Data Quality Assessment and the Situated Nature of 'Best' Research Practices in Biology. *Data Science* 16(32): 1–11.

Leonelli, S. (2018b) Re-Thinking Reproducibility as a Criterion for Research Quality. *Research in the History of Economic Thought and Methodology* 36 (B): 129–46.

Leonelli, S. (2019a) Data – From Objects to Assets. *Nature* 574: 317–21.

Leonelli, S. (2019b) Scientific Agency and Social Scaffolding in Contemporary Data-Intensive Biology. In Wimsatt, W. and Love, A. (eds.), *Beyond the Meme: Articulating Dynamic Structures in Cultural Evolution.* Minneapolis: University of Minnesota Press, 42–63.

Leonelli, S. (2020) Scientific Research and Big Data. In Zalta, E. (ed.), The Stanford Encyclopedia of Philosophy. https://plato.stanford.edu/entries/science-big-data/.

Leonelli, S. (2021) Data Science in Times of Pan(dem)ic. *Harvard Data Science Review* 3(1). https://doi.org/10.1162/99608f92.fbb1bdd6.

Leonelli, S. (2022a) How Data Cross Borders: Globalizing Plant Knowledge through Transnational Data Management and Its Epistemic Economy. In Krige, J. (ed.), *Knowledge Flows in a Global Age: A Transnational Approach.* Chicago: University of Chicago Press, 305–43.

Leonelli, S. (2022b) Process-Sensitive Naming: Trait Descriptors and the Shifting Semantics of Plant (Data) Science. *Philosophy, Theory and Practice in Biology.* 14: 16. https://doi.org/10.3998/ptpbio.3364.

Leonelli, S. and Ankeny, R. A. (2012) Re-Thinking Organisms: The Epistemic Impact of Databases on Model Organism Biology. *Studies in the History and Philosophy of the Biological and Biomedical Sciences* 43(1): 29–36.

Leonelli, S. and Lewandowsky, S. (2023) *The Reproducibility of Research in Flanders: Fact Finding and Recommendations.* Thinkers report of the KVAB (Royal Flemish Academy of Belgium for Science and the Arts).

Levin, N. et al. (2016) How Do Scientists Understand Openness? Exploring the Relationship between Open Science Policies and Research Practice. *Bulletin for Science & Technology Studies* 36: 128–41.

Levin, N. and Leonelli, S. (2017) How Does One 'Open' Science? Questions of Value in Biological Research. *Science, Technology and Human Values* 42(2): 280–305.

Levins, R. (1966) The Strategy of Model Building in Population Biology. *American Scientist* 54: 421–31.

Lewandowski, S. and Bishop, D. (2016) Research Integrity: Don't Let Transparency Damage Science. *Nature* 529: 459–61.

Longino, H. (1990) *Science as Social Knowledge*. Princeton, NJ: Princeton University Press.

Longino, H. (2001) *The Fate of Knowledge*. Princeton, NJ: Princeton University Press.

Longino, H. (2003) *The Fate of Knowledge*. Princeton, NJ: Princeton University Press.

Louafi, S. et al. (2022) Communities of Practice in Crop Diversity Management: From Data to Collaborative Governance. In Williamson, H. and Leonelli, S. (eds.), *Towards Responsible Plant Data Linkage*. Dordrecht: Springer, 273–88.

Love, A. C. (2008) Explaining the Ontogeny of Form: Philosophical Issues. In Plutynski, A. and Sarkar, S. (eds.), *The Blackwell Companion to Philosophy of Biology*. Malden: Blackwell Publishers, 223–47.

Lusk, G. and Elliott, K. C. (2022) Non-epistemic Values and Scientific Assessment: An Adequacy-for-Purpose View. *European Journal for Philosophy of Science* 12: 35. https://doi.org/10.1007/s13194-022-00458-w.

McLeod, M. and Nersessian N. (2013) The Creative Industry of Integrative Systems Biology. *Mind & Society* 12: 35–48.

Mäki, U., Fernández Pinto, M. and Walsh, A. (eds.) (2018) *Scientific Imperialism: Exploring the Boundaries of Interdisciplinarity*. London: Routledge.

Massimi, M. (2022) *Perspectival Realism*. Oxford: Oxford University Press.

Maxson Jones, K., Ankeny, R. A. and Cook-Deegan, R. (2018) The Bermuda Triangle: The Pragmatics, Policies, and Principles for Data Sharing in the History of the Human Genome Project. *J Hist Biol.* 51(4): 693–805.

Mayernik, M. S. (2017) Open Data: Accountability and Transparency. *Big Data and Society* 4(2). https://doi.org/10.1177/2053951717718853.

Merton, R. (1973 (1942)) The Normative Structure of Science. In *The Sociology of Science*. Chicago: University of Chicago Press, 267–78.

Merton, R. (1968) The Matthew Effect in Science. *Science* 159(3810): 56–63.

Miedema, F. (2021) *Open Science: The Very Idea*. Dordrecht: Springer.

Miles, C. (2019) The Combine Will Tell the Truth: On Precision Agriculture and Algorithmic Rationality. *Big Data & Society* 6(1). https://doi.org/10.1177/2053951719849444.

Mirowski, P. (2018) The Future of (Open) Science. *Social Studies of Science* 48(2): 171–203.

Mitchell, S. (2003) *Biological Complexity and Integrative Pluralism*. Cambridge: Cambridge University Press.

Montgomery, L. et al. (2021) *Open Knowledge Institutions: Reinventing Universities*. Cambridge, MA: MIT Press.

Morgan, M. 2010. Travelling Facts. In Howlett, P. and Morgan, M. (eds.), *How Well Do Facts Travel? The Dissemination of Reliable Knowledge*. Cambridge: Cambridge University Press, 36–86.

Morgan, M. (2014) Resituating Knowledge. *Philosophy of Science* 81(5): 1012–24.

National Academies of Sciences, Engineering, and Medicine (2018) *Open Science by Design: Realizing a Vision for 21st Century Research*. Washington, DC: The National Academies Press.

National Academies of Sciences, Engineering, and Medicine (2019) *Reproducibility and Replicability in Science*. Washington, DC: The National Academies Press.

Nature (2018) US Proposal for Defining Gender Has No Basis in Science. *Nature* 563: 5.

Nerlich, B. et al. (2018) *Science and the Politics of Openness: Here Be Monsters*. Manchester: Manchester University Press.

Nersessian, N. (2022) *Interdisciplinarity in the Making*. Cambridge, MA: MIT Press.

Nguyen, C. Thi (2021) Transparency Is Surveillance. *Philosophy and Phenomenological Research* 00: 1–31. https://doi.org/10.1111/phpr.12823.

Nowotny, H. (2021) *In AI We Trust: Power, Illusion and Control of Predictive Algorithms*. Cambridge: Polity Press.

O'Neill, O. (2002) *A Question of Trust: The BBC Reith Lectures 2002*. Cambridge: Cambridge University Press.

Open Letter (2021) Support Data Sharing for COVID-19. COVID-19 Data Portal. www.covid19dataportal.org/support-data-sharing-covid19.

Open Science Collaboration (2015). Estimating the Reproducibility of Psychological Science. *Science*, 349(6251). www.science.org/doi/10.1126/science.aac4716.

Oreskes, N. (2019) *Why Trust Science?* Princeton, NJ: Princeton University Press.

Oreskes, N. and Conway, R. (2010) *Merchants of Doubt*. London: Bloomsbury.

Owen, R., von Schomberg, R. and Macnaghten, P. (2021) An Unfinished journey? Reflections on a Decade of Responsible Research and Innovation. *Journal of Responsible Innovation* 8(2): 217–33.

Popper, K. (2011 (1945)) *The Open Society and Its Enemies*. London: Routledge.

Potochnik, A. (2017) *Idealisation and the Aims of Science*. Chicago: Chicago University Press.

Prainsack, B. (2020) The Political Economy of Digital Data. *Policy Studies* 41(5): 439–46.

Radder, H. (1996) *In and About the World: Philosophical Studies of Science and Technology*. New York: The State University of New York.

Radder, H. (ed.) (2010) *The Commodification of Academic Research*. Pittsburgh, PA: University of Pittsburgh Press.

Radder, H. (2019) *From Commodification to the Common Good*. Pittsburgh, PA: University of Pittsburgh Press.

Romero, F. (2019) Philosophy of Science and the Replicability Crisis. *Philosophical Compass* 14: e12633. https://doi.org/10.1111/phc3.12633.

Ross-Hellauer, T. et al. (2022) Dynamics of Cumulative Advantage and Threats to Equity in Open Science: A Scoping Review. *Royal Society Open Science* 9(1): 211032. https://doi.org/10.1098/rsos.211032.

Rothblatt, S. (1985) The Notion of an Open Scientific Community in Historical Perspective. In Gibbons, M. and Wittrock, B. (eds.) *Science as a Commodity: Threats to the Open Community of Scholars*. Harlow: Longman Publishing Group, 21–76.

Rouse, J. (1987) *Knowledge and Power: Towards a Political Philosophy of Science*. Ithaca, NY: Cornell University Press.

Rouse, J. (2002) *How Scientific Practices Matter*. Chicago: Chicago University Press.

Rouse, J. (2015) *Articulating the World: Conceptual Understanding and the Scientific Image*. Chicago: University of Chicago Press.

Sheehan, N., Leonelli, S. and Botta, F. (2023) From Collection to Analysis: A Comparison of GISAID and the COVID-19 Data Portal. Preprint. https://doi.org/10.1101/2023.05.13.540634.

Solomon, M. (2001) *Social Empiricism*. Cambridge, MA: MIT Press.

Solar, L. et al. (2014) *Science After the Practice Turn in the Philosophy, History and Social Studies of Science*. London: Routledge.

Strasser, B. J. et al. (2018). Citizen Science? Rethinking Science and Public Participation. *Science & Technology Studies* 32(2): 52–76.

Thompson, E. (2022) *Escape from Model Land: How Mathematical Models Can Lead Us Astray and What We Can Do About It*. London: Basic Books.

UNESCO (2021) Recommendation on Open Science. https://en.unesco.org/science-sustainable-future/open-science/recommendation.

United Nations (2019) *Towards Global Open Science: Core Enabler of the UN 2030 Agenda*. https://research.un.org/ld.php?content_id=50970566.

Vermeir, K. et al. (2018) *Global Access to Research Software: The Forgotten Pillar of Open Science Implementation*. A Global Young Academy Report. Halle: Global Young Academy. https://globalyoungacademy.net/activities/global-access-to-research-software/.

Walsh, K. (2018) The Darker Side of Baconianism. Blog, University of Otaga (6 December). https://blogs.otago.ac.nz/emxphi/the-darker-side-of-baconianism/.

Williamson, H. and Leonelli, S. (eds.) (2022) *Towards Responsible Plant Data Linkage: Data Challenges for Agricultural Research and Development*. Cham: Springer.

Williamson, H. F. et al. (2023) Data Management Challenges for Artificial Intelligence in Plant and Agricultural Research. *F1000Research*, 10: 324. https://doi.org/10.12688/f1000research.52204.2.

Wyatt, S. et al. (eds.) (2000) *Technology and In/equality: Questioning the Information Society*. London: Routledge.

Wylie, A. (2003) Why Standpoint Matters. In Figueroa, R. and Harding, S. (eds.), *Science and Other Cultures: Issues in Philosophies of Science and Technology*. New York: Routledge, 26–48.

Wylie, C. (2021) *Preparing Dinosaurs: The Work Behind the Scenes*. Chicago: Chicago University Press.

Acknowledgements

This Element could not have seen the light without the support of five organisations: the European Research Council, which supported my work on Open Science with award 335925 'The Epistemology of Data-Intensive Science' and award 101001145 'A Philosophy of Open Science for Diverse Research Environments'; the University of Exeter, especially my colleagues at the Exeter Centre for the Study of the Life Sciences (Egenis) and the Institute for Data Science and Artificial Intelligence; the Global Young Academy, whose group on Open Science I coordinated between 2014 and 2017, learning enormously from all contributors; the European Commission, who selected me as an advisor on Open Science policy as part of the Open Science Policy Platform in 2016 and the Mutual Learning Exercise on Open Science in 2018, thereby providing me with first-hand experience of the challenges confronted by ongoing reforms; and the Wissenschaftskolleg zu Berlin, whose hospitality and unique intellectual ethos provided an ideal setting for systematizing these ideas during the 2021–2 academic year. I owe a great debt to Wiko staff and fellows, particularly Anya Brockmann, Chris Kelty, Hannah Landecker, Liza Lim, Mark Hauber, Ilya Kliger, Anthony Ossa-Richardson and Alberto Pascual-García; and to Egenis colleagues, including: Chee Wong, who facilitated much of my work over the last decade; brilliant researchers Louise Bezuidenhout, Niccolò Tempini, Nadine Levin and Hugh Williamson; long-term interlocutors Rachel Ankeny, John Dupré, Brian Rappert, Adam Toon and Adrian Currie; the PHIL_OS team, especially Rose Trappes and Paola Castaño, who provided precious comments on an early draft, as well as Alfiya Yermukasheva, Michel Durinx and R. Ankeny. Past and present PhD students, including Sara Green, Stefano Canali, Elis Jones, Alex Mussgung, Arthur Vandervoort, Nathanael Sheehan, Fotis Tsiroukis, Emma Cavazzoni and Joyce Koranteng Acquah, provided invaluable inspiration. I also thank the hundreds of researchers, librarians, information specialists and policy makers with whom I was privileged to discuss Open Science policies and practices over the years, especially James Anthony-Edwards' team in Exeter Parts of Chapter 3 are adapted from "Open Science and Epistemc Diversity: Friends or Foes?" (2022, Philosophy of Science 89(5): 991–1001). I dedicate this work to Michel, whose unwavering support and loving commitment to challenging my every unwarranted assumption is a daily lesson in how life should be lived; and Luna and Leonardo, for whom I hope openness will always be an opportunity and never the status quo.

Cambridge Elements ☰

The Philosophy of Science

Jacob Stegenga
University of Cambridge

Jacob Stegenga is a Reader in the Department of History and Philosophy of Science at the University of Cambridge. He has published widely on fundamental topics in reasoning and rationality and philosophical problems in medicine and biology. Prior to joining Cambridge he taught in the United States and Canada, and he received his PhD from the University of California San Diego.

About the Series

This series of Elements in Philosophy of Science provides an extensive overview of the themes, topics and debates which constitute the philosophy of science. Distinguished specialists provide an up-to-date summary of the results of current research on their topics, as well as offering their own take on those topics and drawing original conclusions.

Cambridge Elements ≡

The Philosophy of Science

Elements in the Series

A full series listing is available at: www.cambridge.org/EPSC

Printed in the United States
by Baker & Taylor Publisher Services

Printed in the United States
by Baker & Taylor Publisher Services